Walter Maclagan Reid

The Culture and Manufacture of Indigo

With a Description of a Planter's Life and Resources

Walter Maclagan Reid

The Culture and Manufacture of Indigo
With a Description of a Planter's Life and Resources

ISBN/EAN: 9783337058791

Printed in Europe, USA, Canada, Australia, Japan

Cover: Foto ©berggeist007 / pixelio.de

More available books at **www.hansebooks.com**

THE

CULTURE AND MANUFACTURE

OF

INDIGO.

PIG STICKING.

THE

CULTURE AND MANUFACTURE

OF

INDIGO;

WITH

*A DESCRIPTION OF A PLANTER'S LIFE AND
RESOURCES.*

BY W. M. REID.

―――――――

ILLUSTRATED:

―――――――

CALCUTTA :

THACKER, SPINK AND CO.

LONDON : W. THACKER AND CO., 87, NEWGATE STREET.

1887.

CALCUTTA :

PRINTED BY THACKER, SPINK AND CO.

CONTENTS.

LIST OF ILLUSTRATIONS.

INTRODUCTORY.

—••+♦♦+••—

IT is proposed, in the following Sketches of Indigo-
life in Tirhoot and Lower Bengal, to give those
at Home and others who have never witnessed
the manufacture of indigo, or seen an indigo factory
in this country, an idea of how the finished, marketable article is produced, together with phases and
incidents of an Indigo Planter's life, such as may
be interesting, instructive, and, it may be, amusing
to friends at Home, and serve as a record of
what an Indigo Planter's usual life now-a-days is.
This book does not profess to deal with the scientific production of indigo, nor the various theories and
processes of late years put forward for the increase
of produce from the plant, or the production of
higher - class colour by chemical means. The old
simple, unadulterated, original process of manufacture is generally adhered to. Planters also, especially " old hands," have a rooted objection to any
" doctoring " stuff; they thoroughly and firmly believe in the real old original " true " blue, with no
admixtures, soda crystals, spirits of ammonia, or
any other potent chemicals.

R., In.   A

This work deals simply with general work and details of an ordinary present-day Indigo Assistant's life, together with a few practical hints which are offered *en passant* for what they are worth ; and it is hoped that this brochure will partly serve to fill a blank long felt, of a record of an Indigo Planter's life, to send Home to friends, or place on the table in India.

In the illustrations the author has endeavoured to give the scenery and natives as they actually exist at this day in the mofussil or more remote districts of India. Native life, habits, and implements are of the very homeliest, and primitive to a degree ; therefore, the reader must not be surprised at some of the modes of agriculture depicted in the sketches.

CHAPTER I.

will suppose the young to-be Indigo Planter already arrived in Calcutta, and his call made on the agents. He has put up at the Great Eastern Hotel for a day or two, eaten his introductory tiffin with the "house," and received his order to proceed up to join his "concern"—somewhere, say, Tirhoot. This will probably be about the 1st of November, when the cold weather has well set in, and the climate is simply lovely. Dates differ as to the commencement of the indigo-season in Lower Bengal and Tirhoot; in the former, the "season" commences from the 1st of October, yearly; in the latter, from the 1st of November. The Indigo Assistant's engagement is yearly, as a general rule, and commences from one or other of these dates, depending on what district he is in. He will go by rail as far as, say, Mozufferpore, from the Howrah Station at Calcutta; *dâks*, that is to say, relays of riding or driving horses or ponies, will be sent out beforehand for him, that he may ride or drive from the railway station to his destination, and a bullock-cart is generally engaged for his luggage. One horse, or

relay, is stationed at about every five miles. Generally, on the occasion of a new hand joining, he is
met by the Manager of the concern he is about to
join, at the station, or by a senior assistant, who will
probably have the pleasure, subsequently, of what
is technically called " putting him through."

The horses or ponies sent out on *dâk* for a new
arrival are not necessarily always the very best which
the " concern " affords. The ordinary Indigo Planter
of commerce is the very soul of generosity, but when
it comes to letting out his best horses on *dâk* for
a friend, it is found that he even has a sneaking
fondness for the general fitness of things, and prefers to suit the animal to the emergencies of the
journey to be performed. The roads are excellent
roads, *as* roads ; but when a planter or the department on that particular lay for the time being, goes
and repairs a road with two feet and a half of loose
earth and large clods just before a long season of
heavy rains, the effect in and about the months of
October and November may be imagined, but cannot adequately be described. Whether riding or
driving *dâks* are sent for our young friend, he will
have to be most particularly careful to keep his
" heye " open about mounting and riding some of
the former. They have been out once or twice on
dâk before, notwithstanding he has not, and they

PLANTING HORSE.

know a trick or two, and he must not be surprised at somewhat violent and exciting starts—full speed reckoning as the very lowest order of velocity, graduating from that upwards, to the sensation of being shot out of a rocket—with an amount of "backing and filling," which would satisfy even a Yankee skipper trying to get up the river Hooghly without the aid of steam.

A favourite way of starting the old *dâk gharry* horses which, in times now long gone by, used to transport the traveller all over India, was to have no nonsense at all about it. When the animal *sat down*, and obstinately "refused" to go, the mild Hindu attendants with characteristic Oriental indifference to suffering, quietly, but promptly, collected straw, sticks, and dry grass, and *lit a fire* under the wretched animal. Under these circumstances the difficulty about starting did not admit of any further argument, and was at once got rid of. If in a mail cart—'one horse in the shafts, and occasionally an auxiliary one rigged stun-sail-boom-fashion '—instead of a *dâk gharry*, one had to hold on pretty tight when the horse did start, which he generally did at one bound, and without giving the least previous notice, even by so much as a cock of the ear. Although some of the factory hacks are difficult to manage both in harness and saddle, it will hardly be found necessary, except in extreme cases, to proceed to such lengths as this last to induce them to " proceed." Another favourite " dodge " was to tie a thong or rope tightly round the pastern of an unwilling starter, and apply vigorous tugs and wrenches to this particularly tender part, until the poor animal had no choice but between immediate starting and a prospective raw leg for a month from the friction of the rope.

Besides the ordinary saddle-nag, the bamboo-cart is most generally used for travelling ; the sketch furnishes the details of one of these vehicles. They *can* be made from part of an old indigo-chest and four pieces of bamboo, and some of them are so made for very rough-and-particularly-tumble work, inasmuch as the wheels holding out, they cannot break. But some are made quite elegant little conveyances, with very little of the bamboo at all about them except the shafts. The real *bonâ fide* bamboo-cart, with plain bamboo shafts and pincer springs, is simply unbreakable, capsize or no capsize, so long as the wheels are good and the two or three screw fastenings required are tightly braced up. They rank with the bullock-hackery, which has, since the British occupancy of India, been unanimously pronounced, by both amateurs and professionals, military and civil, mercantile, agricultural, and marine (many a gallant sea-captain in Cook or Brown & Co.'s

III.

H&P

ASSISTANT'S BAMBOO CART AND NAG.

Thacker, Spink, & Co., Calcutta.

jauntiest hired buggy who has tried it), to be absolutely invulnerable. The more elegant so-called bamboo - carts are generally rather gingerbread affairs,
which require constant repairs. They are apparently
made up of putty and paint at some of the larger
up-country stations, and come to pieces during the
rains, or before they have been in use a twelve-month·

CHAPTER II.

THE FACTORY-ASSISTANT'S BUNGALOW, ETC.

I will suppose our young friend to have at last, after the usual journey in trap or saddle, arrived at the "Sudder" or head factory, the residence of the Manager of the "concern." He will probably be entertained here for a few days prior to being despatched to his out-factory, where he will take up his residence, I will hope, for some years, gradually qualifying himself to become some day a manager himself.

The Assistant's bungalow, or house, will probably be much like the illustration given. His cook-room will always be at some little distance from the house; it has been found well that this, in an Eastern clime, should be so There are weird legends extant, of matutinal coffee strained through the garment which covers the foot Have not puddings been boiled in banians and pocket-hanky-thingumajigs lent by the bearer from—not amongst the clean things ? Is not hare (hair) soup served up regularly until the *bobar-chee's* (cook's) ambrosial locks are perforce seized and shorn by the stern order of the *sahib*, who—with all his love of game soup and pie, to say nothing of curry—holds, in this case, with the fastidious man in

Assistant's Bungalow.

Back Verandah.

Bath Room

Dining & General Room

Bed Room

Dressing Room

Pantry   Front   Verandah

Store Room

Interior of Cook Room in India

Thacker, Spink, & Co., Calcutta.

the backwoods restaurant, who preferred his flies and molasses in separate plates, and likes his "hair," if it must come to table, placed in a prominent place, say, on a separate dish of the largest size, and sent thereon, with best compliments, back to the *chef* in the cook-room.

The bungalow will probably contain a dining-room, bedroom, dressing and bath rooms, with a pantry and "godown" (or store-closet), with back and front verandahs, as shown in the plan given. Nothing, in this particular instance, is sacrificed to art. This amount of accommodation is all that is really needed for a bachelor ; it doesn't matter one bit if one doorway isn't exactly opposite another ; in fact, it is rather an advantage than otherwise, as an Aryan brother who has a complaint of a peculiarly noxious or distressing character, such as wanting to borrow money on personal security or anything of that sort, simply reconnoitres until he finds a place in the garden or compound, from which angle—if under a convenient tree, so much the better—he can distinctly see through the house so that, move which way the Assistant will, the, "grievancer's" eye is upon him. In vain the unfortunate aspirant for indigo honours tries to evade the steadfast eyes of the watcher — smoke as he may, read as he may, intently as he may dive his nose into the page or newspaper—his eye will slowly, gradually, but surely, be dragged, line by line, letter by letter, up the page to the top of the column and over the edge, and his eye will inevitably meet the eye of the "watcher" fixed patiently upon him. Animal magnetism! It's no use; he KNOWS you MUST look at him; he never stirs, he never speaks; he sits there, mute, asking for nothing, seeking nothing. Oh, no! Wishing for nothing. Only, you will have to, and MUST, eventually, LOOK AT HIM; and herein lies his triumph! And, the instant his eye catches yours, there flashes out upon his face, as from a lighthouse on the coast, suddenly such an expression of abject misery and woe, want and injured innocence, as no words ·but mute appeals can do justice to, dying away again as suddenly the instant your eye is removed from him. In this case the human eye and animal magnetism has great and unconquerable effects, and it has even been known to attract a boot-jack, half a brick, or similar heavy bodies with great velocity to within an ace of the "watcher's" head, accompanied with much low and violent rumbling, as of bad language and ill-temper of a furiously aggressive but suppressed kind. Therefore, the more irregularly one's doors are "let in," so much more is the "silent watcher" checked off.

At the most, the company our Assistant is likely to see will be, occasionally—say, once or twice a month —a neighbouring assistant whose cultivation adjoins his own, who may drop in for tiffin (or the mid-day meal), and perhaps a visit once a week or so from his Manager just to see how he is getting on, and with whom he most likely returns to the Sudder factory for the night as a pleasant change to the monotony and dulness of the out-factory. If his Manager is a family man, he may hear a few tunes on the piano, and a few of the Old Country songs which will remind him of those at Home ; and he will return to his lonely bungalow the next morning the better for the pleasant evening, one of the few he is likely to pass until the cold season comes round again, then he can, perhaps, get away to the races, or to the Behar Volunteer Cavalry parades, or to Calcutta for a few days, where his numerous wants will inevitably lead him to China Bazar, there to be assailed, by native brokers, hawkers, and middle-men clinging to his gharry with cries of 'tin-box,' 'orginette,' 'meershum pipe,' 'portmantoo,' 'esleepin-suit,' 'ready med close,' 'photogripes taken,' etc.

CHAPTER III.

GENERAL LIFE, AMUSEMENTS, ETC.

THE general life of an Indigo Assistant is much as follows, and varies but little throughout the hot and rainy months; but during the colder months, *viz.*, November, December, January, and February, more festivity prevails,—races, hunting, and hockey meets are held at different stations; the Behar Mounted Rifles hold cold-weather parades frequently,—and the Assistant's life is, thus, much enlivened.

As a rule, the Assistant, in ordinary work, rises at 5-30 A. M. ; he then takes *chota hazree* (or little breakfast), consisting of a cup of tea, a slice of toast, and a couple of eggs, and his horse is brought round meanwhile ready saddled. He starts out immediately after swallowing his tea to inspect his lands, the ride extending sometimes to eight, nine, or ten miles, not at a gallop by any means, but at a steady inspecting pace. He notes all the work going on in the different parts of his charge; perhaps meets his Manager, also out on his round; gives and receives the news; issues his orders; and returns about 11 A.M. He has a good bath, and, quite refreshed, looks over his letters, perhaps, or writes one or two, then comes tiffin, or the mofussil mid-day meal, which is, in fact, the principal meal of the day. If he gets a newspaper, after tiffin, he sits and cons it over a comfortable smoke, and, perhaps, in the hot weather lies down for a couple of hours under the punkah. At 3 p. m. or so he rises, and then most of the writing business of the day is got through. The factory "writer" comes with the books and accounts, and they are examined ; the items written out, opposite the vernacular, in English by the Assistant, and the amounts totalled and balances checked, and the books initialed or signed. After this follows a confab, long or short, with the *gomashta jemadar* or headman of the factory, who takes his orders from the Assistant, and some pointed censure also sometimes, about the work present and to come. By this time evening is closing in, and the Assistant may walk or ride out again to see some work close at hand, or may look up the factory cattle, and see how they are being fed and generally cared for, or he may take a stroll round the stables and garden, and then in to dinner.

This somewhat monotonous life to some is frequently enlivened by a run with the " bobbery pack." The " bobbery pack " proper is usually a joint-stock affair—the property jointly of two or three assistants

of a somewhat sporting character—and consists of
dogs of every breed, colour, and shade possible, from
the squab little bow-legged, half-bred, cur doggie to
the no less half-bred, lanky, big-jointed cross between a Rampore and a pariah dog, or a greyhound
and a kangaroo dog ; these last, by the way, being
the finest dogs for India. Notice being sent round,
each member contributes his *chiens de chasse, boule-
dogue, settare, kangarow,* or *griound,* as the case may
be, and then, hurrah ! for *le sport.*

Mr. Jackal's haunts are pretty well known, and
towards the evening is the best time to find him, just
as he is issuing forth for his nightly rambles. The
mehters, or dogmen, follow in various stages of entan-
glement with their charges. When, at last, the scent
of wily "jack" is hit upon by some cur doggie of
more than ordinary nasal intelligence, and after much
slipping of collars bodily over dogs' heads, and thereby half strangling some of the most eager, and a
variety of contrivances in the way of leashes loosened,
away streams the crowd, yelping, yahouping, yelling,
and barking, each one at the top of his throat, some
before, some behind, all howling their level best ;
the *griounds* and *kangarows* leaping like bony race-
horses gone mad amongst a flock of sheep; the smaller
dogs rolling and tumbling over in their eagerness,
but scrambling up, and going in again—barking

vociferously all the time, though, even when on the broad of their backs ; each one determined, if possible, not to be quite the very last " in at the kill." At last Master Jackal gains a patch of pretty tall indigo, and there remains *perdu* for a time. The noble pack are at fault, but do their best ; the little terriers as they come up and big dogs—each jumping up as high as they can from the ground—peering about for a glimpse of the " game." Some of the little terriers actually bouncing about like indiarubber balls, glancing wistfully here and there, till, at last, Mr. J. having got his wind again, slinks slyly out and makes for some favoured haunt, replying to the yell with which his appearance is greeted by a whisk of his tail only, and a single defiant glance backwards over his shoulder.

But he has miscalculated his enemies. With mighty sweeps of his hind legs—like a race-horse at his best, or the revolutions of a fly-wheel of a steam engine at full pressure—and with gleaming eyes and cruel fangs, an enormous half-bred kangaroo dog, the king of the pack, flashes down upon him. In vain Mr. Jackal dodges and runs his hardest, his enemy closely pursues him, and is backed by the two best greyhound pups in the pack, each almost a match already, single-handed, for a Jackal, and fine promising young dogs. It is all up with Master Jackal

now. The Assistants check their steeds, and pause
to witness the end, which is fast approaching and
which will round off Master Jackal's days, poor fel-
low, right in that little bare patch of cultivated soil
glistening now a short way ahead of his labouring
feet. Ha! the big dog has him down ; a sideward
turn and one savage despairing snap of the teeth,
but never a cry, are all the Jackal's defence, and in a
white puff of dust, shot high into the sultry air, dogs
and jackal roll over together, and present a confused
mass of lugging and tearing legs and teeth, hot chok-
ing throats and fierce flaming eyes, till the body of
yet another "jack" lies bleeding and lifeless, torn
and mangled past recognition, on the bosom of its
mother earth.

This amusement may be varied by an occasional
shooting excursion, or a race or hockey meet, or even
by worrying a cat in the reservoir ; but the latter
amusement for a dull afternoon will probably strike
the reader as being quite too refined and intellectual
to suggest itself to every one ; yet there are some
manly natures whose innate sporting instincts will
lead them to the pursuit of "game" under any difficulties.

The Assistant will find that, with a little trouble,
and very little expense, he can make his factory-
home quite a snug little place. With everything

coming from Home now at such cheap rates, a few really artistic, first-class oleographs or chromolithographic landscapes, animals, and horses' and dogs' heads, etc., after Landseer ; photographs of celebrated pictures and people, or what not, may easily be got for a moderate price for his rooms. Flowers may be had in any quantity, as almost every factory has a garden. Cage-birds in India are plentiful, and of many beautiful varieties ; some can be taught to speak, whistle, or sing to perfection. A good aquarium is also a source of endless study and delight to those scientifically inclined or those who have a taste for natural history, as there are an immense variety of extraordinary specimens of fish, aquatic animals, insects and plants, etc., to be got from the Indian rivers, streams, ponds, tanks, and ditches. Altogether there are very many worse countries than India to live in, though certainly for some months in the hot weather it is very trying. But let the Assistant devise some means of indoor employment, improvement, and amusement. He can study the reading and writing of the language if inclined, and will find such study invaluable to him as a planter in afterlife, or he may sketch in water or oil colours and adorn his domicile with the results. The Assistant may carpenter a bit in the way of fretwork brackets, or turnery, all of which will help to pass the dull times away and

serve to adorn the bungalow. The flower garden will
also require a good deal of care, and will furnish
employment for every spare minute. Then there is
always work of some sort to do, and should our
Assistant still find the time of an evening hang heav-
ily upon his hands, he should get some instrument,
never mind of what sort, from a hundred-guinea
violin or piano down to a penny tin whistle, and
practise it until he can play. He need never fear
the sound jarring on the native nerves, or offending
native ears: these are the only ones likely to hear
him. The native of India is positively *fond* of dis-
cordant sounds ; and he will probably, on slipping out
into his verandah in the darkness of the stilly night,
after performing every direst kind of excruciating
discord on his instrument for the time being—violin for
preference, if any—find, as one assistant did, his *chow-
kidar* behind the door, wrapped in profoundest ex-
tacies at the dulcet sounds he had been producing ;
and he will be greeted with a respectful but confiden-
tially whispered : " *Wah ! wah ! Huzoor ! ap hhoob
bajanewallah hai*," which being interpreted is Pushtoo-
Sanskrit for : " Well and neatly executed, oh ! pre-
server of the poor ; Paganini Sahib himself, were he
here, might have been willing, or even anxious, to quote
the ever-verdant Artemus, to die after having heard
such a masterly medley of sweet sounds as thou, oh !

distributor of unbounded charity to unworthy reci-
pients, hast produced ! " Seriously speaking, it is
often a great thing for one of the gentlemen present
at a gathering to be able to play even a few dances
to amuse the company. It enlivens the evening and
pleases many, besides giving the ladies and those who
are " dancing " gentlemen an opportunity for a dance,
to which the former, as a rule, never object. Even to
be able to play the Highland Fling well on a penny
whistle is an accomplishment, and a good accomplishment too at a " meet ; " and to be able to knock out a
couple of jigs and a hornpipe or so on an old fiddle,
will often serve to enliven a dull evening in the rains
and set the flow of conversation going again. The
very squeak of a fiddle resembles Mr. Punch's well-
known call note of " R-r-r-r-r-r-rooooooo—titoooit "
in its suggestion of fun. And in that dire calamity
which doth sometimes befall the male portion of the
human race, *viz.*, the absence of ladies and the piano,
much may be got out of gentlemen amateur perform-
ances. Even a periodical Christy Minstrel gathering would not be impossible, did the district number
amongst its Assistants a " banjo," " fiddle,"" tambou-
rine " and the indispensable " bones ; " the corner-
joking-and-singing-and-conundrum " Massa Jonsons "
being also forthcoming. What dreadful and unchid-
den jokes might not one make on one's Manager,

disguised as " Massa Benson ;" what cunning allusions
to paucity of salary, or cutting references to the
scantiness of one neighbour's crop might not one
make under cover of Massa Jonson's "mechanical
wit" and darkened countenance!

•

CHAPTER IV.

THE MANAGER.

THE Manager is appointed by the proprietor or joint proprietors of the concern—*i.e.*, the head factory with all its outworks—under a general power-of-attorney of the fullest nature, to superintend and have a care over all and every interest in the said concern. In proportion as his power is full, so is his responsibility heavy. The efficient management of a large, or indeed of any, indigo concern, far from being a sinecure, is attended with many difficulties, and demands the exercise of an amount of judgment and sagacity not perhaps required in any other profession in India. The Manager has virtually to be a " jack of all trades," whilst the comparative mastery which he has attained over most of them is to be admired. His duties are multiform, and his labour unceasing. The routine of his work is everlasting. To him the treadmill is indeed a daily round, but it is the mental as well as the physical one ; a well-balanced mind is equally necessary with a robust frame and sturdy limbs : *mens sanâ in corpore sano* is an indispensable requisite in the good Manager. We have said that his duties are many, and we will now proceed to describe some of them for the incipient assistant, the manager of the future.

The Manager's chief work is in his *katchary* or office, and his chief duty the regulation of all matters intimately connected with the cultivation and manufacture of indigo. His other business, though arduous, and often more vexatious than his business proper, necessarily holds a secondary place, being created by, and therefore tributary to, it. In most head factories the account-books of all the factories are examined and passed every month by the Manager, who approves, or otherwise, as he deems fit. The Assistant, although enjoying a certain amount of discretionary power, is not master of his own expenditure, being responsible to the Manager by whom he is appointed ; whilst the Manager is directly responsible, not only for his own, but also for the acts of his subordinates of every grade. Thus, a vast deal of trouble is cast upon the shoulders of the Manager of the concern. By factory-accounts I would be understood to mean all such as pertain to the actual working and carrying on of the factory itself in all its branches. The ramifications are numerous. We have the establishment, European and native ; all accounts connected with the cultivation of the *ziraths*, or fields, in which indigo is directly and entirely sown by the factory, *sua propria*—those which have to do with *ryothi* or such lands as are cultivated by tenants under advance from the factory. Then, there are the accounts which deal with charges for repairing old and constructing new buildings or works : the large item of cattle demands a close and unflagging scrutiny ; whilst such items as go-down stores and general charges, under which categories come all stores purchased during the year for consumption or use, and all and sundry— the thousand and one incidents ever cropping up in a factory ; *mahai* or manufacturing charges and machinery charges, which accounts embrace all the multitudinous items accompanying and result-ing from the manufacture of the plant into dye ; and the working, regulating, and keeping in order of the machinery. These, and a few others, I would designate the *accounts proper* of a concern, all of which come under the inspection, demand the attention, and require the sanction, of the Manager. Thus, it may be seen that, even were the Manager's labours confined within the limits we have detailed, his office work would not be easy, nor his *katchary* a " Castle of Indolence."

But he is not allowed to rest here. No sooner has he escaped from the Liliputians than he is attacked in force, and captured by the Brobdingnagians. They seize him in the form of *putwari's*, or village accounts. He feels weary and inclined to resist ; but they are too strong and too pressing, and he must plod on. These accounts arise in this way. To acquire land in which to carry on his industry the planter has to—or rather has hitherto had to, for this is all being changed now—enter into treaty with the surrounding *zemindars*, or landholders, for leases of their villages. These villages being apportioned out to *ryots*, or tenants, the *zemindar* is unable to treat for any special portion or division of a village, and, in consequence, the planter is forced to take the whole, with all its advantages or drawbacks. The advantages he anticipates are the obtaining of lands in which to cultivate indigo ; the drawbacks are the *patwari* and his accounts. These are sent him for his sins. We will deal afterwards with the question of *zemindari* and the feature in it called *tikhadari*, or leasing. Meanwhile, back to the Manager, and his *patwari's* accounts : these consist of *siahas* (vouchers), *jamma bundies* (rent-roll), and *jamma kharchas* (debit and credit account). The *siahas* accompany all moneys paid in by the *patwari* as village-collections; they may come daily or weekly, according to the varying customs prevailing ; indeed there is scarcely any regulating rule, each Manager having his own pet system. Over these troublesome little papers, which he must sign and confirm, the manager spends a weary time, testing, inquiring, doubting, believing, and sometimes, whispering soft nothings into the *patwari's* ear. These papers are checked by a book called the *kistibundi* (instalment-book) and the daily cash register. The *jamma bundies* and *jamma kharchas*, alas! come together; but "the wind is tempered to the shorn lamb," and they only come up for searching inquiry once a year. These give the total rental of the village, showing the *khast* or holding of each tenant with its rental; the amount received in cash and that still to be collected; the impracticable lands from which no rent accrues; the tenants, real and imaginary, who have disappeared from the village during the year, carrying the whole year's rent with them (the *patwari* invariably sets them down as delinquents to the full); the cesses for which the village is liable — fisheries, tanneries, etc. The *patwari*, foul bird, has preyed the year out upon the pecuniary vitals of the tenants, but now he dons his brightest plumage, and it is indeed difficult to detect his obscene nature; but duty is duty, and the Manager has a high and enduring sense of it. So, into the *patwari's* papers—not only his own, but those of the out-divisions as well — the Manager goes, consenting and dissenting, cutting and clipping, doubting and badgering, until he arrives at a final settlement with his worst enemy—the *patwari*. All this time the Manager is enveloped in a dense cloud of falsehood and chicanery, which he disperses, sometimes with advantage to himself, but, too often, with profit to the peculating *patwari*. These village-accounts are like the stars, eternal; but unlike them, gloomy and dark—offering little light to the Manager, except such as his own skill and science may evolve. But the intricacies of this fell branch of a Manager's work the Assistant must find out for himself; I offer merely a descriptive outline, which the space at my command forbids me to enlarge.

The Manager has also to do the ordering and guiding of his Assistants as to all outdoor work; and although, as a rule, injunction and obedience are, in this case, so to say, synonymous terms, yet it is worrying. Generally, the Manager prefers the morning, before going out, for writing his subdivisional letters and orders, English and vernacular. These are not few. All insubordination, all mismanagement or misbehaviour of native subordinates, all truculencies of tenants, all matters referring to law or to land, and everything dependent upon changes in the weather, are referred to him for orders. He is the *alpha* and *omega* of the factory-system—the sun, round which all the lesser bodies move.

Correspondence with agents and Home proprietors is a task which few managers like, although performed with fair regularity. The Calcutta letters and the weekly letter Home are *bêtes noirs*, but the Manager looks upon them as part of his duty, and they are written. In correspondence with the agents, the Manager reports on the progress of the "concern," arranges for his seed, and orders all stores required during the current season. He keeps the Home proprietor *en courant* with the state of affairs generally, and of lawsuits and *mamlahs* (negotiations) particularly ; to the agents he yearly sends his estimate for the coming year's outlay, which they either approve on receipt or forward Home for orders. Once approved, the wherewithal for the ensuing year is steadily supplied by the agents, to whom monthly abstracts of expenditure are forwarded.

Another branch, still, of the Manager's office work, is the conduct of suits at law ; some managers have more to do in this way than others. Some factories are comparatively free of these pests, whilst in others the Manager's time and mind are both much exercised by them. The immunity or otherwise from such suits seems mainly to depend upon locality, and whether the *zemindars* and *ryots* have acquired a taste for litigation or not. Be this as it may, no manager is quite exempt. The factory *mokadama* (lawsuit) throbs along in three great arteries — civil, criminal, and revenue. In the Civil Courts, the Manager prosecutes the bond-bound *zemindar*, the swindling carter under a five years' agreement, the defaulting servant, and the peculating *patwari ;* or, occasionally, the tables are turned upon him by all four parties. For the immediate supervision of such cases he employs an ingenuous gentleman called a *vakil* (native pleader), who takes fees. Criminally, he also wages war upon the perjured bond-bound *zemindar*, the similarly perjured carter under a five years' agreement, the defaulting servant, and the peculating *patwari,* when he can "run in" these latter. Two other new patients present themselves for legal treatment—the trespasser and the thief. At this tribunal the interests of the Manager are usually protected by that spotless specimen, the *muktar* (pleader, minor). In the Revenue Department, presided over by the civil and small cause judge, all suits relating to renewal of leases from *zemindars*, realising of outstanding balances from *ryots*, boundary disputes, rights of occupancy, enhancement, distraint, etc., *ad infinitum*, drag their weary length along "with ever lengthening chain ; " for, is not a suit before a *munsiff* (judge of small causes), the first cousin of Jarndyce *v.* Jarndyce? These suits are also in the tender care of the *vakil.* Sometimes a case of either of the two first descriptions goes up to the High Court, but this is not an every-day occurrence, and a manager may have one or two during his whole term of management. The revenue-cases go to the Commissioner or Board of Revenue, and the fiat of the latter is final.

I will close my description of the office functions of the Manager with a cursory glance at that section of his work which is called in the planters' "shop" language "doing *mamlahs*" (*i.e.*, negotiations) for leases of land and villages. Here the Manager comes into direct contact with the *zemindar*, the latter generally waiting upon the former at his factory. Indeed, it is in most cases the impecuniosity of the *zemindar* which leads him to seek the Manager, whom, in his heart, he considers only a shade better than the *mahajun* (usurer). But his needs are urgent, and the *mahajun* presses hard upon his flank. The factory is his last resort. The usurer's rates of interest are iniquitous; the Manager's demands are fair, if his terms are hard. Early and late the *zemindar* betakes himself to the factory, and the Manager receives him in his *katchary*. Then ensues a time of hard swearing over "cooked" papers on the one part, and protestations and much unbelief on the other :

"My village is the fairest field for indigo in Behar. Where can you find such lands and such rates?"

" True, my friend, your village is good and your lands are fair ; but you may well ask where such rates can be found—they are extortionate."

" You misunderstand ; I want to imply that, for the quality of the soil, the rates are low, too low. I shall be ruined if I conclude with you on your hard terms.   I shall go back to the *mahajun.*"

As you please, my friend ; I do not particularly want your village.   I but deemed I did you a favour. Good-bye ! "—Mutual grasping of hands ; exit *zemindar.*

This kind of mummery goes on for some time ; the *zemindar* is proud and greedy, and the Manager, who is dying for the village, would not show anxiety for worlds. "He will come again," he says to himself ; and he is right : here he comes. His demands are considerably less this time, and the Manager, in a deprecating sort of way, agrees to consider his proposals.   The rent-roll of the village he has seen *ad nauseam.*   He has it all by heart — that heart which he has set upon the village.

The *zemindar* proceeds : " 'The *jamma bundi* of the village is so much ; profits, deducting all expenses, so much ; you will advance me so much money at eight per cent ; in so many years at this interest, the rent-roll of the village will liquidate your money, principal and interest ; and, at the end of the term,

you will pay me cash so much—the balance of rent
after liquidation—and return me my village, paying
me 10 annas rent for your *ziraths* for ensuing year.
There, sahib ! see what a friend of yours I am ; and,
ah me ! see how long I am out of my own ! My loved
ones at home will die ; it is their homestead, and
they curse me for alienating it."

The Manager shakes his head and looks unspeakable
things. What would his proprietors say ; what would
other managers say, if they heard he had done such
a *mamlah ?*

" No, no, no !" the *zemindar* gasps. He did think
his present proposal was glittering enough. Well,
he would throw himself upon the mercy of the
mulak malik (lord of earth). What would he
have ?

Then the Manager, and his looks are stern and
high : " I acknowledged your *jamma bundi—that*
we have settled. I will advance you rupees   at
12 per cent., absurdly low, but out of courtesy to you.
In   years my principal and interest are liquidated
from the income of your village, with a balance due to
me of   rupees, which you will pay on demand.
Your village will not then be relinquished, but will
continue in my possession for   years more on a
simple lease. You understand ? These are my terms,
to which our long friendship constrains me."

Alas, alas! what can he do, the slouth-hound behind, this dark ford before! He slowly, reluctantly passes the ford and is safe. He closes with the Manager's terms.

This is what is called in Behar the *sudowah putowah mamlah*, which, though we have placed first, managers like worst; where they see their money gradually ebbing away, until the village is handed back to the *zemindar*, who promptly asks for more.

But the methods of managers in treaty with the *zemindar* are many and variform. There is the *khusk tikha* (simple lease), by which the village is farmed for a term, the Manager paying the *zemindar* the stipulated rental in instalments; the *zarpeshgi* lease, where the money advanced constitutes the principal, and the annual rent of the village, the interest. Under this form of *mamlah* the village is never relinquished until the principal is paid up in full, and it obtains much favour with the Managers of Behar. There is the *mukarrari*, which is a simple lease in perpetuity. On lands leased in this latter way, vats and all factory-buildings are erected. We need hardly say that large villages or areas are seldom obtainable in this way. There are many other changes rung on the everlasting *mamlah;* but those mentioned are the chief phases in which it confronts the Manager.

The Manager of a concern has also a good deal of riding about. He is regular in his morning visits to the *zillah* (out-cultivation), ordering, countermanding, and inspecting work generally. Indeed, there is no Assistant busier than he. But, especially in large concerns, the *zillah* work is often looked after by a personal Assistant employed for that purpose. For a description of *zillah* work, *vide* Chapter III, page 12 ; the Manager's and Assistant's work in this department being essentially the same. Outworks are now and again looked up, and, in this way, a pleasant relaxation from the toils of *katchary* obtained. To be stern, yet conciliating, firm, yet pleasant ; to know when to hold and when to yield ; to be courteous and honourable—never departing from his word once pledged ; to have an intimate knowledge of his profession, and a good deal of *savoir faire,* together with a good knowledge of business : in fine, to know how to " become all things to all men, so that he may win " something—these are some of the attributes essential to the making of a good indigo manager.

———

R., In.   C

CHAPTER V.

THE AMLAHS.

THE *amlahs*, or native staff of a factory, vary in number and responsibility almost as much as they do in integrity and capability. At a large head factory the *amlahs* are many, but as the tyro will not come much in contact with some of these during the years of his probation, I will pass them over, merely noticing those *amlahs* who are indispensable to the outwork, and common to head factory and outwork alike.

First in the list comes the *jemadar*, a sort of native personal assistant to the Assistant, equivalent to the English bailiff or Scottish grieve. On this functionary devolves the arranging, ordering, and inspecting of all out-door work ; he is the *deus ex machinâ* of the *zillah*. The cultivation of lands is pre-eminently his province, whilst the choosing of new, and realising of old and worn out lands falls within his jurisdiction, subject to the approval of the Assistant. In a well-ordered outwork, the duties of *jemadar* should not be to hire any extension of power in collateral directions,—this being an inducement to dishonest transactions, which the worthy *jemadar* can rarely, if ever, withstand ; and, indeed, the young Assistant will, as a rule, find it difficult to curb the *jemadar*, even when confined to his own proper walk. When first put in charge of an outwork, the Assistant will necessarily be very much under the tutorship (we had almost said at the mercy) of his leading *amlahs*. But the latter term is perhaps too sweeping, because, if the Assistant is active and anxious to penetrate all the intricacies of his work, he will never consider it a task of supererogation to personally test and satisfy himself upon all the matters connected with, and reported upon by, the *jemadar*. His own common sense will teach him that such constant investigation must be so regulated, that the *jemadar* may never dream that there is any suspicion connected therewith. Not infrequently the *jemadar* of long service, regarding the young Assistant as his pupil, and, being a native, the sequence is, that he puts on a certain amount of swagger which, though the newcomer from England may not recognise as such, his subordinates in the factory are fully alive to, and they behave to the *jemadar* in proportion to the amount of " side " he can put on. The behaviour of the *jemadar*— even to the softly attuned tones of his voice when speaking—is obsequious to the extreme to the Manager. To that grave and great of all the factories, he does not speak above a whisper, while to the Assistant his behaviour is, in many cases, aggravatingly insolent, in such a way as to render it difficult to the young hand to curb this tendency to " cheek." In some factories in Behar, where there is not always an Assistant in charge, the *jemadar* who acts in that capacity loses his head, and the next Assistant finds he has a most difficult task. Generally the remedy is drastic, and the *jemadar* gets turned out. If the Manager objects to this, and does not even substitute some punital equivalent for the " sack," the Assistant who still remains does so as a nonentity, and is not worth the salary he gets. But I am glad to certify that the Manager seldom or never lends a deaf ear to the Assistant's grievance, nor is he slow to afford redress.

The *lalla*, or *munshi* as he styles himself—aping the title of the real *munshi*, who resides at the head-factory, and who works some of the legal and other wheels of the concern—is the most important servant the Assistant has. He looks after the accounts of the outwork.

He is sometimes hand-and-glove with the *jemadar* ; the *nazri lalla* forming the corner-stone of a confederacy rich in loot to the worthy partners. When the parties are not on terms of amicable commerce, there is a certain amount of friction which the Assistant must overlook or watch silently ; then the acute youngster is initiated into various tricks and secrets of the gentle Hindu which, in the innocence of his heart, he never dreamt of before. He must, however, be deaf to mutual recriminations, keeping both ears and eyes wide open, always remembering that, of all the members of the conspiring clique, the *munshi*—by caste a *kaisth* or writer—is the least to be trusted and most to be guarded against. Having once come to the righteous conclusion that there are none good amongst them, not even one, the Assistant will proceed to deliberate upon their respective degrees in the guild of villainy, and whatever rank the other may hold, the *lalla* will be *facilè princeps*, the grand master. His are all the " tips " for dexterous fudging of accounts ; grievances of every kind pour in, but never pass the threshold of the *munshi's* office until the petitioner has paid his footing. Indeed I know of factories where the *munshi* himself was inaccessible, until the *peon* at the door had received his footing.

In fine, the *munshi*, or native accountant of an outwork, is a power for evil, and it is only by constant, though secret, surveillance that the Assistant will be in any way able to check those extortionary instincts which are the very marrow of his bones. As a rule his house and family are far away ; and the Assistant would do well to remember that the golden

rule of the *munshi* is to make the largest possible
"pile" in the least possible time.

Next in position comes the *hazri nawis*, or *lalla*,
who is in charge of the home *ziraths* and the ordinary humdrum work within the factory precincts—
such as repairs, including carpenters', blacksmiths',
masons'—and their coolie helpers—and other general
work common enough for an indigo factory. This
amlah arranges, counts, and keeps a register of all
coolies, ploughs, factory ploughmen, carters, chaff-
cutters, carpenters, blacksmiths, masons, etc., all
people in receipt of daily wage or monthly pay.
This daily register must be scrupulously tested by
the Assistant in person. I have always found it
best to accompany the *hazri nawis* in his morning and afternoon rounds, counting and dotting
down in a pocket-book the attendance of all grades
with him twice a day. The object of a second
hazri in the afternoon is to see that all those
counted in the morning are again present in the
afternoon — an hour's respite being given to the
work-people from twelve to one, for their mid-day
meal. The *hazri nawis's* pickings are considerably
less than those of either *jemadar* or *munshi*, and
are pretty much confined to "*dasturi*"—an infini-
tismal blackmail levied by him upon coolies, which
they give voluntarily, or, at any rate, in consideration

of their being allowed to work when they like, and
exempted when home-work is to be done ; and a
small sum monthly from factory carters, ploughmen,
etc., on receipt of their pay. These petty pecula-
tions it would perhaps be as well that the Assistant
did not notice, unless expressly brought before him :
they are recognised institutions of the country of
his exile—indeed, *dasturi* explains itself, being an
adaption of the word *dastur*, custom—to meet the
exigencies of such anomalous partnership as exists
between the *lalla* and his coolies. The Godown
factory stores are often in charge of the *hazri nawis*,
to whom is given a weighman, who is generally a
bunnia by caste.

The smaller fry who make up the scant contingent
of *amlahs* at an outwork, *zilladars* and *tokdars* may
be disposed of with a short notice. The *zilladar* is
a kind of *jemadar's* mate, who has under him the
supervision of a portion of the *zillah* and three or
four *tokdars*. I am free to confess I never much liked
zilladars: they seem to be *de trop*, especially at an
outwork, and are an arrangement by which the *jemadar* baulks his *zillahs*. From them he often takes his
reports at second-hand, and bravely serves them up
to the Assistant as the result of his own inspection.
Nor are they in favour either with *tokdars* or *ryots*,
from both of whom, too, they extort their quota

which takes the various forms of *salami*, or loot.
Of a truth these middle-men are not worth their keep.,

The *tokdars* are in charge of sections of the *zillah*,
which may range in extent from 30 or 40 to 150 or
200 *bighas ;* such sections are called *toks*, literally
divisions, and are cultivated, sown, and the plant cut
by hired labour on the *ticca* (jobwork) system, the
tokdars receiving advances, at fixed rates, from the
factory. These men are directly under the *jamadar*,
and are in receipt of small pay, so that they find it
necessary to eke out their existence by petty extor-
tions of all kinds. A favourite dodge with the *tokdar*
is to measure his work with a large *luggi*—measuring
rod ; thus making the work done by the hired labourer
come out considerably less than, by the proper factory *luggi*, it actually is. He sometimes contrives it
so that his " mate " turns up late of a morning with
the weeding or other coolies, when the wretches are
made to work double-tides, but are informed in the
evening that they shall expiate their fault of the
morning by receiving half wage. The factory-rates
being generally pretty liberal, the *tokdar*, in the above
and many other ways, succeeds in at least doubling
his pay in any month.

The youngster would do well to remember that
all natives are most correct and acute judges of
character. Indeed, it is not too much to say that

whilst he is still groping in darkness as to the disposition of his several *amlahs*, they have within a very few days taken his measure pretty correctly ; and even as their estimate is, so will their behaviour and general demeanour be. He must learn that any display of preference is a fatal weakness, which the favoured one will lose no time in working upon to his own advantage, whilst the less fortunate ones will go about their work in a sulking, slovenly way, the result of the ascendency asserted over them by the favourite.

The *amlahs* are a necessary evil—a veritable thorn in the flesh to the hardworking, conscientious Assistant ; but such as they are, he cannot get on without them. He will find, as he becomes more initiated into their tortuous ways and more conversant with the manner of their lying, that even they might have been worse, and he congratulates himself that his *amlahs* are not as those of other men. The abyss must be deep indeed where there is not still a deeper depth, and the Assistant finds that the *amlahs*, if not exactly angels of light, are human beings after all.

CHAPTER VI.

CONSTRUCTION AND TESTING OF DRILLS.

THE indigo-drill is a curious mechanism, and, from its appearance, might arrogate a high antiquity, although, in reality, quite as recent in Behar as indigo itself, of which it is the offspring. Moreover, it lays claim to first cousinship with agricultural implements for the same purpose, and, some of them, of exquisite mechanism, in Europe. In *Gleanings of Science*, No. 5, for May 1829, there is a description, with a drawing, of the drill plough of Tirhoot used in the cultivation of Indigo, the advantages of which are thus stated. " The land intended to be sown having been previously rendered fine by repeated ploughings and harrowings, with the common native plough and bango, and all clods, weeds, etc., removed, the trough filled with well-dried seed to about the level of the iron axle, and everything being adjusted, the bullocks are urged on. The shares cut the furrow, the wheels of the machine turn those of the trough, the slanting holes bored in the wheels of the trough, during their passage through the seed, take up each one or more seeds (seldom more than one), and, in the downward part of their revolution, unload themselves with precision into the hoppers, which lead them into the hollow of the ploughshares, which last deposit the seed in the furrow and inclose the seed in an instant."

Let me attempt a description of the indigo-drill, as seen in most factories. Of late years several patent drills have been invented and tried, but none of them seem to have established a reputation equal to that of the original. The construction of the implement is as follows : On 2 inches to 2½ inches diameter wheels, 3 inches thick, invariably made of shisham wood, and roughly tyred with common hoop-iron, is erected a trough, 4 feet by 1 foot, inside measurement, of the same material as the wheels, but generally, now-a-days, lined with 16 inches to 18 inches B. W. gauge galvanised iron ; the average depth of this trough being 6 inches, *i.e.*, from beneath axle (1 inch square bar) passing through centre of trough, and up to level of which seed must only be filled. The side walls of the trough are 2 feet long by 2 inches thick, also made of shisham. It is requisite that all the above-named parts be made of shisham (the oak of India). We have next to describe the front-board or platform in which are inserted the nipples for sowing the seed, and also, in front of these, the shares, by which the earth is opened to receive the same. This is also 4 feet long, inside measurement, by 9 inches to

10 inches broad, by 3 inches thick, again of shisham
wood, and to it the trough is attached.   This platform
is bored at distances of 6 inches apart in a double
line to receive the tubes and shares, the former of
which, beginning flush with the platform, extend 9
inches down inside the shares, which are generally
16 inches in length, including a 2-inch point or tip.
A simple board, 1 inch thick, is fixed in on end
between the nipples and shares to prevent the seed
from being thrown outside.   Thus, in a drill of 4
feet length, inside, there are eight tubes and eight
shares.   The tubes are made of B. W. gauge galvanised iron ; are round and gradually tapering towards
the bottom, being generally ¾ inches round at top.
The shares are merely flat bar iron, 1 inch by
¾ inch, upon which, at bottom, pieces of ⅛ inch sheet-
iron, gently graduating outwards, both to receive the
tube end and also to make a furrow, are welded on.
These share sides are usually 4½ inches to 5 inches
broad, and 2 inches wide at outer rim.   Inside the
trough, and upon the axle, with which they revolve,
are placed, at equal distances with tubes and shares,
small round wheels, 7 inches in diameter, which
are slightly perforated round the disc with small
round holes, which, with every revolution, raise the
seed in the trough and eject it into the tubes in
the front board or platform.   The average number

of holes in a 7-foot diameter seed-wheel would be 16.

It now only remains to describe the shaft. This is a thin piece of wood 9 feet long by 2 inches thick, sloping, when the drill rests upon the ground, from 4 feet from ground at point to 9 inches where it joins the platform ; here it is dovetailed in, and secured on either side by strong stanchions of wood or iron, the former for preference   The object of giving the oblique direction from point of shaft to platform of drill is in order that the drill may adapt itself to bullocks, who pull by means of a yoke.   Bullocks being of all sizes, slots are made in the outer end of the shaft, so that thus the drill is heightened or lowered to suit the size of animal employed ; it being of the last importance that the shares should be kept at an uniform depth so as not to endanger the germination of the seed.

Now that I have described the construction of this primitive implement, I will proceed to see how it is to be tested for sowing the indigo.   I would here note that I have described what is known as the eight-shared drill.   There are, however, many sizes, ranging from the four-shared to the twenty-shared, pulled by either single or double pairs of cattle.   I have merely selected the above as what I consider the most useful size ; and I may add that all factory-made drills are of similar build.

Seed is sown so many *seers* to the *bigha* (roughly, ¾ acre) by the little seed holes in the discs before mentioned. Now, the drill being of wood, wood being perishable, the Indian climate variable and severe upon this article, and the doctrine of absolute certainty in the sowings being a first principle with the planter, it will be seen that these little seed holes, tested religiously and laboriously this year, must not be trusted the next. It, therefore, becomes a necessity that the drills should be tested. This process, excruciatingly dear to the heart of every Assistant, comes into requisition about the middle of February of each year. There are two different systems now in use—the old and the new—the first rapidly becoming shouldered out by the last. By the old system—we will suppose you wanted to sow at the rate of twelve *seers* per *bigha:* a small component portion of a *bigha* was measured off in a square, lined off; then, having filled your drill trough and yoked your cattle, round they went until the ground apportioned off was drilled. I must mention that the seed, before entering the drill, had been weighed ; all the test seed went into the ground. The cattle being unyoked, the balance of seed in the trough was re-weighed, the difference between first and second weighments being what remained in the plot. This difference gives the ground-work of the simple calculation in proportion—if so much land takes so much seed, how much will so much land take? And round went the drill, and the little seed lobes were manipulated with red-hot iron and shellac until the required proportion had been attained.

This circuitous and vexatious means to an end is now fast fading into a reminiscence. The second method deserves more attention as being more scientific and, at the same time, more correct. The number of revolutions of wheels of the different diameters in a *bigha* being obtained, and the small seed wheels being properly adjusted to throw equally and, at the same time, a certain amount per *bigha;* the calculation is again made as to how much seed should be thrown in fifty revolutions to obtain the desired average per *bigha;* all becomes plain sailing then. The drills are all collected in the factory compound, and the testing begins. A level plank is placed under the drill to be tested, which has all its shares put in as if ready for sowing ; small bags of indigo cloth are then attached to each tube, one spoke of the wheel is marked with chalk to certify each revolution, and round goes the wheel by hand, the revolving agent being a small stick inserted between the spokes and firmly grasped in both hands. On the completion of the fifty revolutions, the little bags are opened out, and the contents of each bag measured in vessels made for the purpose—the best measure is a medicine glass—and the little seed holes are again pierced or filled up until the required average is attained.

In conclusion, I would remark that the presence of the Assistant is quite necessary to obtain honest testing.

———

IV.

(TUMMED) DIGGING LANDS FOR INDIGO.

CHAPTER VII.

CULTIVATION, MANURING, MEASURING, AND IRRIGATION.

THE system of cultivation, sowing, cutting, and manufacturing the indigo plant, being precisely the same in both the Sudder and out-factories, except that the Sudder, being larger factories, now-a-days generally use steam-power for beating and pumping purposes, one description will do for both.

The first process in cultivation which our Assistant will have to give his attention to is the *tumnee* or digging, which begins immediately after the manufacturing for the preceding year ends. This digging is simply the backbone of the sowings, and if badly and carelessly done, there is little hope for the future crops. It is done by many men in a line (as in the illustration given), each armed with a *kodal*, or native spade, and the whole lot presided over and watched, not only by native peons and *burkundazes* (or stick men), but also, very frequently, by the Assistant himself. The stick men carry large bamboo staves poking out a long way from the shoulder behind them ; they also curve their moustachois most Bombastes Furioso-usly, but are very harmless individuals after all, and not by any means so fierce as they appear.

On this work of digging deep down and well depends, in a great measure, the future health and growth of the indigo plant, and it cannot be too carefully supervised. A gang of *badmashee* coolies —from *badmash*, a rogue or rascal—will so scamp the work if not thoroughly well looked after, that, though the land will look, on the surface, to be well and properly dug, it will in reality be hardly dug up at all; and if the Assistant will dismount and scrape away the top-soil, he will find that the men have hardly dug four inches in depth, whereas they should have dug to the full depth of the *kodal*, (or digging implement used). If an assistant gets his *tumnee* well done, the rest of his cultivation is comparatively easy, but if an assistant's *tumnee* is scamped over in the first instance, it will be a never-ending and constant source of worry and anxiety to him, and his future cultivation will never look thoroughly well done; and, as to manuring it afterwards, if necessary, it is only throwing half the manure and labour away to attempt it.

I may here mention that, in the English or Irish acceptation of the term, a spade is an implement unknown to the native of India; he cannot with his bare feet use it, for though the sole of his foot is

PLOUGHING INDIGO LANDS.

tough enough, it would hurt it too much, and, then, his muscular power is, as a rule, not sufficient to enable him to use it with anything like as much effect as a European could. It is difficult to imagine to what use a native could or would put a spade if supplied to him ; probably he would utilize it in some way of his own undreamt of by its owner or giver, as some coolies did when supplied by an energetic railway contractor with wheelbarrows to accelerate the throwing up of some earthwork. They proceeded to carry them loaded with earth on their heads instead of the light baskets they had been using, thus considerably retarding instead of forwarding the work in hand. The only use a native could make of an English spade would be, perhaps, to denude it of its wooden handle, using the same for fuel, and bake *chapattees*, or unleavened cakes, on the flat of it, as the Australians used to fry bacon in the first rush to the diggings.

The *tumnee* or digging ended, the ploughing next begins, which partly stirs up and breaks the clods. The ploughs are mostly of native construction, and made of wood, with an iron ploughshare let in, though sometimes iron ploughs of an English pattern, adapted to bullock-draught, are used. , The ploughs generally go in a string of five, six, or more, following one another at short distances, and each a little to the left of the preceding—one set ploughing, say, from north to south, and another set working at right angles across their furrow from east to west.

After the land has been well turned over and the size of the clods somewhat reduced, a roller is passed over the ground to further pulverize the earth. These rollers are either large blocks of wood carved or cut with a thread like a screw, or smooth cylinders of stone, and are drawn by a pair or two of bullocks or buffaloes, usually the oldest and most worn-out animals in the factory, which not being smart and active enough for ploughing, are put to this slow and arduous work, in which dead weight tells more than strength and activity. After rolling down the clods, the ground is again ploughed up, perhaps three or four times, according to the dryness, stubbornness, or clayiness of the soil. The smaller clods which remain are then finally broken by hand by gangs of, from 50 to 100 women and children, boys and girls, in one long row, who keep up a perpetual din, beating time on the clods with thick short sticks amidst clouds of dust until there is not a lump left bigger than an ordinary walnut. In February the sowings commence, and seed drills are busily at work. After the drill follows an instrument, drawn by bullocks, like a long bamboo ladder, which smooths the earth over the seeds, and then the lands are left. In from four to five

CLEANING WEEDS AND BREAKING UP CLODS.

Thacker, Spink, & Co., Calcutta.

ROLLING LANDS, BREAKING CLODS.

Thacker, Spink, & Co., Calcutta

VI.

Spreading and Digging in (Seetee) Manure.

days the seeds germinate, generally on or about the third day, and the two first leaves of the plant peep forth.

As the lands become exhausted year, by year, a portion of them is taken up and trench-manured with *seet*, which is the refuse plant after the indigo is extracted. This refuse is carted out and thrown in large heaps in regular lines and intervals over the land, and dug-in in lines or trenches and well covered over by the *kodal* or native spade. All the refuse indigo plant of a factory which is taken out of the vats after steeping, is, however, not used in this way, owing to a large proportion of it—at least half at each factory —being set aside, to be dried as fuel, and then stacked near the boiler for future use in manufacturing the coming season's indigo. It is, in fact, a valuable commodity both as manure and fuel, and is guarded from theft, as far as possible, by the planters, though a good deal of it is " sneaked " by the poorer classes of natives employed about the factory during the cold months to warm themselves with at night; for blankets and warm covering are scarce with them, and fuel also.

This system of manuring is quite unknown in Lower Bengal, where the alluvial deposit from the yearly overflow of the Ganges takes its place and renders it unnecessary, for it would in all probability be washed away. But further up-country where these great yearly inundations do not occur, it is both necessary and imperative to supply the exhausted lands with some equivalent in the shape of manure, and for this *seet* is the best and cheapest.

After the young plant appears, the measuring of the crop is commenced, and this is performed with the assistance of a wheel, exactly one *cottah* in circumference, twenty of which *cottahs* go to a *bigha* in length. Two wheels are used,—one working the length of the field, the other the breadth. The wheel is furnished with a piece of iron protruding from the edge, which at each revolution strikes against a small brass bell or gong attached to the framework and notes off the number of revolutions.

An average of the quantity of plant apparent is then taken, and noted in the margin of the measuring-book. A full crop is put down as a sixteen-anna crop, that is, sixteen annas in the rupee, or full measure ; a three-quarter crop as twelve annas ; a half crop as eight annas, and a bad show of plants as four annas, *i.e.*, a quarter crop. This latter will, in all probability, have to be broken up and re-sown as soon as possible. Many planters when engaged in the measuring have their tiffin sent out to them, and enjoy it sitting in the shade under a tree : it saves all the trouble of going home and returning in the middle of the day, often a good distance, and, as the days are

MEASURING INDIGO FIELDS.
Thacker, Spink, & Co., Calcutta.

short, the time can ill be spared. By taking a gun many a good shot has been had at unexpected stray heads of game, such as a chance hare, leopard, civet cat, floriken, wisp of teal or plover, deer, wolf, fox, or what not. They do shoot foxes in India, generally for the fur, which is very handsome in the cold weather. It is often the case that when without a gun many good opportunities are lost—as the American farmer said when he saw for the first time a couple of young " dudes " perambulating the principal avenue of the city. " Gosh ! " exclaimed the rugged cultivator of the only free soil in the world, " what things a man DOES see *when he harn't got his gun* along with un." It is always advisable to have the " iron " handy, and a couple of cartridges in one's pocket.

Often in very dry weather irrigation becomes necessary, or much of the plant would die for want of moisture. There are several modes of raising water in vogue with the natives and planters, *viz.*, the slung basket, worked by two men at a time ; the lever and bucket, balanced with a lump of mud at the end of the pole, as shown in illustration ; the pair of bullocks and inclined plane ; the China pumps, the Persian wheel, and the various steam pumps ; but the three latter sorts are too expensive for most native cultivators. Irrigating indigo is unknown in Lower Bengal, and is only practised in the Upper Provinces.

CHAPTER VIII.

WEEDING, CUTTING, CARTING, MANUFACTURING.

AFTER the plant has well shown itself it will require to be well weeded, and this must be carefully done, or much of the young plant will be trodden under foot and destroyed. The weeding instrument is of the shape given in the illustration, and is used by the natives in a squatting position. It is driven down to the root of the weed, and a sharp push is given forward which cuts the weed off below the surface, about two inches deep. Care must be taken in doing this, or much of the young plants' roots may also be injured. The weeds should also be collected upon the edge of the field if possible, and not left in heaps on the field, as each heap will cover and kill a certain number of young and healthy plants, and, say, at ten heaps to an acre, each heap destroying a hundred plants, the loss must be considerable.

The plant is ripe and fit to cut from about the end of June, earlier or later, according to the weather. It is carted *as soon as possible* after cutting to the vats and packed in the steepers as per illustration. After steeping in water for some twelve or thirteen hours,

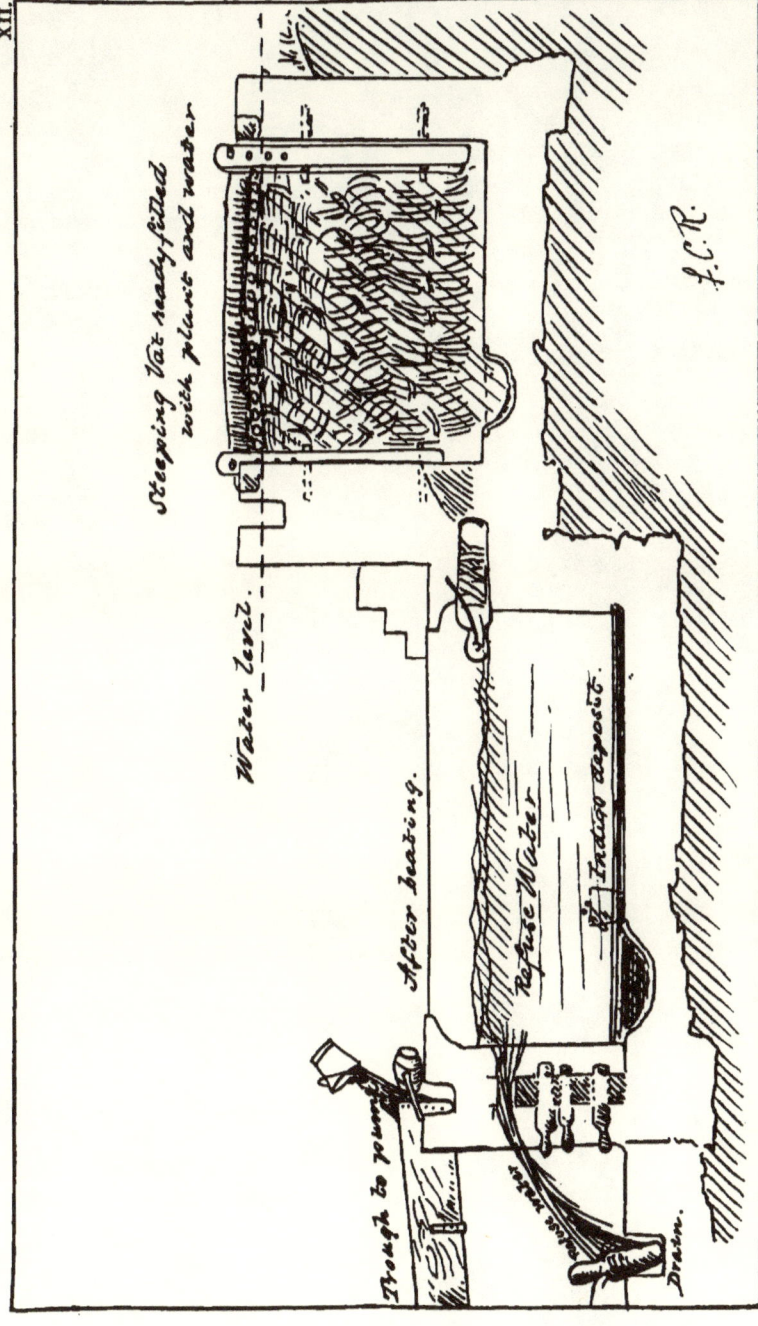

Steeping Vat ready filled
with plant and water

Water level.

After heating.

Refuse Water

ⁿᵈ Indigo deposit.

Trough to pump

Drawn.

f.C.R.

INDIGO VATS.

PUMPING UP THE INDIGO.

according to the time required for fermentation, the liquor is drawn off into the beating vat, and beaten or thrown up into the air as much as possible with wooden paddles or staves, by ten or twelve men in two rows of six each, beating towards each other inwards. The sides of the beating vats are curved inward at the top to prevent the liquor splashing out. Steam machinery beating or paddle - wheels are now used in very many factories for doing this. The beating occupies about from two to two and a half hours or a little longer, according to the weather being fine, warm, and sunshiny, or cold, dull, and wet.

After beating, during which the liquor changes from bright to dark green, then to purple, and then to dark purple-blue, when the indigo grain is formed, the grain is allowed to settle. The beaters run round inside the beating vat several times to give the liquor a rotatory movement, and then jump quickly out, leaving the indigo in the vat to settle to the bottom. When this is accomplished, in two or three hours, the bad or refuse water is run off, as shown in the sketch, and only the indigo remains at the bottom of the vat. This is then washed down by two men, and lifted out of the vat by buckets and drained into the reservoir under the boiler pump. With this pump, the indigo is then pumped up into the boiler (see sketch) and *heated to boiling point ;** it is then run out of the boiler on to the " table " to settle (see sketch), and all that runs out with the water through the table sheet is at once pumped back into the sheet again (see same sketch) until it runs off perfectly clear, leaving the indigo on the sheet with a small quantity of water in it, which must be removed by pressure. The sheet is now doubled carefully over the indigo, and weights put on to press it ; much of the water is thus got rid of, and when no more will run, it is uncovered and lifted in buckets or square wooden vessels into the screw presses (see illustration). Here it is put into square boxes or presses, well and neatly lined with pressing cloth, the boxes themselves having holes bored in them

---

* *Copper* boilers were mostly used in the old Bengal indigo factories, as they are much superior for indigo boiling purposes to iron boilers ; they are thinner than the iron-plate ones, and transmit the slightest heat readily, which is a great point, as indigo is less likely to burn and get blackened, and can be gradually raised to boiling point without overdoing it. Copper, also, does not corrode with damp nor flake off under the action of fire. They are more expensive than iron boilers, much more so, but the money is well laid out on them : they last for many years, a less quantity of fuel is required than with a thick iron boiler, and the heat can be regulated with a thin copper boiler which cannot be done with a thick iron plate boiler, inasmuch as the latter sometimes gets nearly red-hot and almost fires the indigo when the firing has been careless and excessive. Every instant indigo is over boiled, it loses quality in colour to a proportionate extent.

XIV.

Indigo on the Table after being Boiled.

XV.

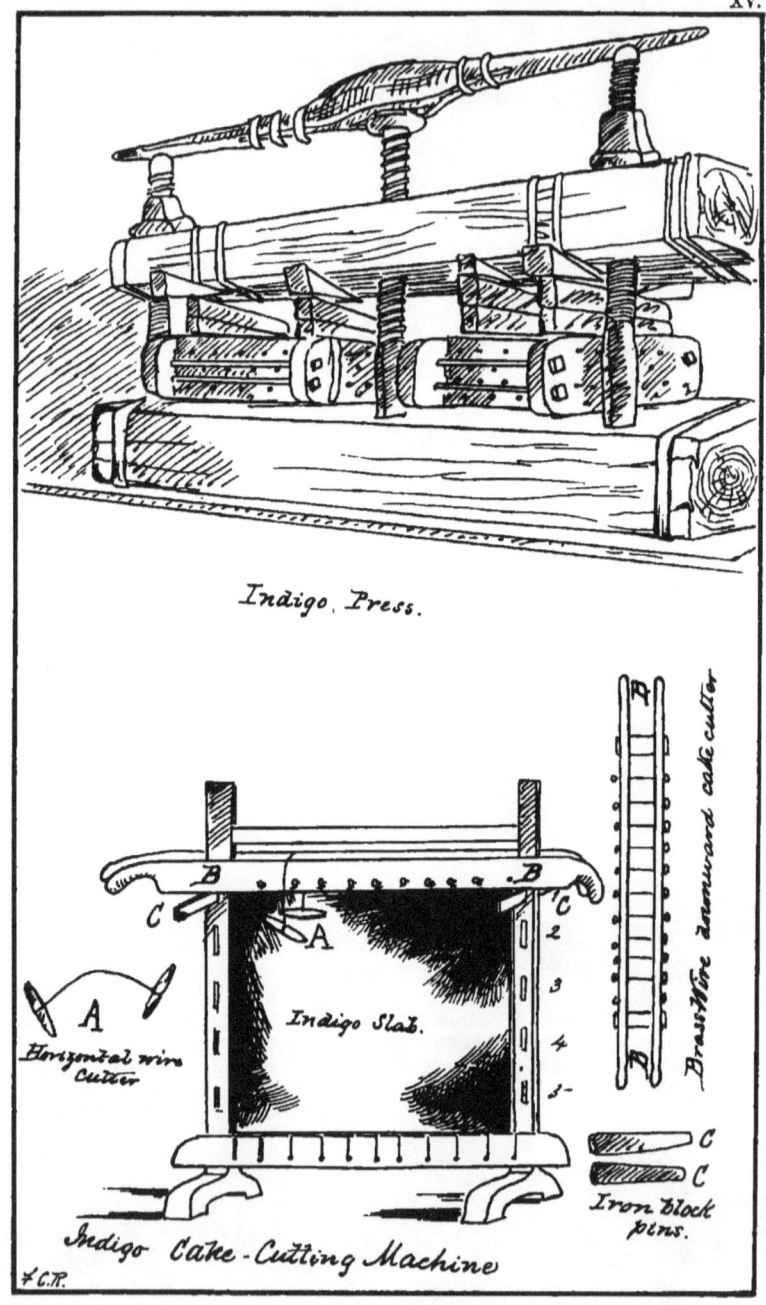

Indigo Press.

Horizontal wire Cutter

Indigo Slab.

Brass Wire drawward cake cutter

Iron block pins.

Indigo Cake-Cutting Machine

in regular rows to allow of the water to drain away, and separating into four pieces each, with a top and bottom, also pierced with holes.

These boxes or presses are firmly clamped and screwed togethe rwith iron screw ties and nuts, and after being filled with the soft indigo, the cloth is carefully and neatly folded down, the top fitted on, and a gentle pressure—just a turn or two of the screw—is applied. As the indigo hardens and becomes of a firmer consistency, a much more powerful pressure is put on until the indigo assumes the solidity and consistency of hard bar soap, in fact, it looks, when cut into cakes, like deep-blue bar soap, and is of about the same density. The indigo is now cut into cakes of about 3¼ inches square, more or less, according to the thickness of the slab,—the thickness of the cake varies. If much water was in the indigo when it "went to press," the cakes will be much thinner than if the indigo was firm and thick and nearly free from water. When it is properly drained free from water, before being put into the screw press, it should allow of being taken up in large masses in the hand without dropping ; when it is very bad, and will not settle and drain off properly under weights, it can only be taken up in a large wooden or copper spoon used for the purpose, many sizes larger than the largest soup-ladle. If, when the indigo is "sloppy,"

it be partially pressed, and the press-box re-opened
and more sloppy indigo put in and mixed up with it, and
the whole re-pressed, each cake will be of a pretty good
size, but each cake will also infallibly be in several
pieces on drying. \The half-pressed indigo will not mix
with the freshly put in indigo, however much it may be
stirred about, and the only effect of much stirring it
about will be that the future cakes will in proportion
be the more in pieces when they are dried. If the
indigo is watery and is half pressed, and the press
then re-opened and fresh indigo filled in, in order to
make large and square cakes, and the fresh indigo is
not stirred up, then each future cake will inevitably
dry in two distinct parts, an upper and a lower one,
differing in proportion in exact accordance with the
quality of each sort of indigo—the first pressed, and
the afterwards added portion in the press box.

It has been found better never to re-open a press
when once filled, but to press it steadily down, and
if the cakes are a little thinner than usual, it cannot
be helped—they will at least dry whole, or most part
of them will, and they will be hard and firm, and dry
" square " and with but slight symptoms of " abdo-
minal contraction."

The reason for the *māl* or *fecula* not settling—as
sometimes happens—giving great trouble, too, when
run out on the table—is generally held to be bad,

stale, or half-putrid water, used for steeping the plant, as the indigo which will not settle on the table is invariably of a most offensive odour, and almost always acts in this way when bad water has unavoidably to be used.

The cure, and the *only* cure, for it, I believe, known, is to press the indigo with gunny-cloth, instead of the usual canvas-like pressing cloth generally used. I have tried this, and found it answer perfectly ; the coarse texture of the gunny-cloth (seed-bags, opened out and well washed, will do) allows the water to escape freely, while completely retaining the indigo grain and the method has only to be tried by those in difficulty with bad, unsettling *māl*, to be appreciated. The only piece of pressing cloth necessary in the operation is a small square piece at each corner, laid neatly inside the gunny-cloth ; the top ends and sides are then neatly folded over in the press-box, the cover placed, and a gentle pressure put on, when the water (perhaps slightly discoloured) flows freely off, gradually causing the indigo to consolidate and admit of firmer pressure being put on till the indigo slab is as hard as may be. I have pressed indigo in this way with common seed-bags (free from rents and holes) which would not settle on the table, and was slushy as pea-soup, and made cakes of it equal to the rest — firm, consistent, and which dried almost perfectly square, with only the slightest symptoms of concavity in the side. The water I was using at the time was taken from a stagnant branch of a stream or shallow inlet in the rains connected with the Ganges, and which, at the time of using, had been stagnant for months, and used by the natives of the village on its banks for all kinds of bathing and washing purposes, and the indigo made with it had a most horrible smell ; but on the Ganges rising and fresh water coming in, these symptoms in the manufacture disappeared entirely and the produce increased, while the indigo made from the clean water settled beautifully, and the usual pressing cloth was again employed.

In attempting to press sloppy *māl* with the usual indigo pressing cloth, the water is so intimately, agglomerated with the indigo itself that it will not run out—the close texture of the wet cloth prevents its doing so ; whereas the gunny-cloth fibre retains the indigo matter, but allows the watery particles to flow off freely. In using the ordinary cloth, it simply bursts under the extra pressure employed in endeavouring to force the water out, and fresh cloth has to be put in at every refilling of the presses ; besides which, when the cloth bursts, the indigo is shot or squirted all over the press-house, escaping under such pressure through the rents in the cloth,

and much loss of indigo is thus sustained, to say nothing of the expenditure of good cloth burst and spoilt.

The cakes are cut in machines specially made for the purpose. The slabs of indigo are put in some machines upright, in some flat, and cut by hand with a brass wire. Each cake is then carefully stamped with the name of the factory and the number of the boiling, to check off afterwards the quality and date of the colour. They are then taken into the godown or drying house, and placed on mat or bamboo trellis-work on shelves in tiers, one above another ; each day's manufacture having its ticket attached to the tier on which its cakes lay, so that by reading the number and remarks on the wooden ticket any specimen or cake of any date can in a moment be at once got down. When the cakes are thoroughly dry they are carefully taken down and cleaned with soft brushes—or, now-a-days, by a machine which has been invented for the purpose by Messrs. Begg & Co., of Calcutta, and they are then packed.

PACKING AND DESPATCHING.

The packing is done with the utmost care and precision, in order to avoid broken cakes and pieces. After all the cakes have been taken down from the shelves and brushed and cleaned, they are packed

according to the boiling and colour list sent up by
the Calcutta agents ; each cake is carefully fitted in
close to its neighbour in rows in the box or chest,
so as to avoid shaking and consequent damage to
the dry cakes.   Native masons are generally, if not
always, employed for this process, as they are more
accustomed to handling that kind of work in building and bricklaying.

A large average of pieces in the packing is to be
most strenuously avoided, as it is a very " bad mark "
on a factory, and argues carelessness and want of
supervision.

After packing, the chests are carefully weighed
and the contents noted ; they are then closed, screwed
down, and re-trimmed all over, and marked ; they
are then again re-weighed and a close calculation
made of actual contents, and are despatched by boat
or cart to the railway station, and thence by rail to
Calcutta.

The wood should be mango at least a year old ;
and the boxes should be kept for one year, as the
effect of using unseasoned boxes is to cause a difference in the weighment at factory, at Calcutta, and,
again, at London.

The tare of the boxes should be always carefully
ascertained by the Manager or his European Assistant.

First.—Previous to commencing operations for the season, have the factory and all apparatus thereunto belonging, put into efficient working order, that the brick and mortar repairs, in particular, may have sufficient time to dry before use.

Second.—The plant being ready for cutting, which may most safely be known by the flowers making their appearance on the plant,—have the reservoirs, steepers, beaters, and everything in the manufacturing house well scrubbed, washed, and cleaned; try the pumps; fit all the pins; dust out and arrange the drying house for receiving the cakes, by having the open mats placed smoothly on the *machans*, and have others placed underneath, to receive the pieces that may be broken off the cakes in turning them or may fall between. Look to your boats or *hackeries* (as may be) for bringing in the plant, and that your coolies and others are ready to work when called upon to do so; not forgetting that all your *challies* (bamboo frames for keeping down the plant under beams) and wood-work for the vats, are properly placed and fitted for use.

Third.—Everything being ready, appoint a day to begin working and apprize your *gomasta* of the same, that he may have also the coolies, *hackeries*, boats, &c., in readiness.

Fourth.—The day having arrived, and everything

R., In.   B

being in readiness ; begin to fill your reservoir at
day-light, that the water may have time to settle
before being required for use ;—and this it will be
well to attend to during the whole of the manufacturing season ;—it certainly being best to use clear
settled water when you can get it ;—bearing in mind,
that as often as you perceive the bottom of the
reservoir dirty and covered with mud and sand, it will
be necessary to clean it.   During the day, as the plant
comes in, measure and stack it up in the shade that
you may be able to load your steeper all at once in the
afternoon, so as to have all done, in this way, before
dark,—taking care in filling the vats, that the water
will run freely out of the steepers into the beaters
when the pins are taken out, by forming a channel
with the plant at the bottom next to the wall.   Much
has been said by experimentalists about placing the
plant in the steepers in particular ways, which, after
all, is not of material consequence, and is ill repaid
by the loss of time in attending to it.   I repeat, stow
the vats so as to let the impregnated water run off
freely, keep the plant horizontally square as you fill up,
that the *chailies* may set square down upon it, and the
beams pin down into the same number hole in the
stanchions.   Thus, if the first beam sets down to the
fourth hole from the top, the rest are to be pinned
down to the same number ; this will give the advantage

of having the whole equally covered with water and thereby aid the fermentation, which is enough. However, you should always have about six inches clear from the top of the steeper, to allow room for the rising of the water undergoing fermentation, otherwise during that process it will run over and waste.

Fifth.—The night being moderately fair, the plant good and not inundated before, in nine or ten hours it will generally be ready to draw off. Now, see that your people are ready, and have the beaters thoroughly washed out before use. Those who aim at making extraordinarily fine blue, steep a much less time, and generally lose a considerable portion of their produce, and not unfrequently make their indigo of too light a colour for general use. As the fermentation goes on, air bubbles rise upon the surface of the water, until at length the vat has the appearance of a broccoli head ; now watch carefully the ebullition of the vat, and the moment it begins to subside and sink down in the steeper, let off the liquid, and you will generally be safe. If it runs out a bright straw-colour, tinged with green, the indigo will be fine ; if a strong madeira colour, good ; if of a very pale straw-green, violet, and dirty-red, bad (coppery.) The first indicates good steeping ; the second, a little too long ; the third, not enough ; and the last, overdone. This will be your guidance,

attending, of course, to the state of the weather, to
steep longer or shorter, according to the quality of
indigo you wish to make.

Sixth.—As soon as the steeper is drawn off into
the beater, send in the coolies. At first beat slowly,
either with bamboo paddles, wooden *fowras*, or the
hands,—the latter way is the best, but more tedious,—
and increase as the froth disappears ; but at no time
beat violently for fear of breaking the grain, and
carefully watch until the whole of the froth disappears, and the vat assumes a dark-blue colour
bordering on black. Then, with the dipper, take up a
little of the matter from the bottom of the beater,
and run it into a white plate. If the grain readily
settles at the bottom, having a clear smooth edge
along the water, it has been beaten enough ; if rough,
and if it does not settle at once, beat more until it does.
Then order the people to throw the contents of six
earthen *calsies* of *chunam* water into the vat, to prevent putrefaction afterwards in the indigo ; and let
them walk a dozen times round the vat and come out.

Seventh—In two or three hours after the beating
is over, according to the state of the weather, the
indigo will have settled to the bottom of the vat ;
then take out the uppermost pin, and go on gradually
down, until the whole of the refuse water is drawn
out, when the bottom of the vat will appear covered

with the fecula. During the time of running off the water have the drain leading to the reservoir where the pump is placed, thoroughly cleaned, after which let the Indigo run into it through the proper strainers.

Eighth.—The fecula being in the reservoir, have it pumped into the boiler through the strainers at the mouth of the pump and at the head of the boiler. Then light the fire underneath, and keep stirring the boiler to prevent indigo from settling and sticking to the bottom. When it comes to the boil, continue it gently so for three hours, stirring occasionally the whole time, during which period get the table well cleaned and washed down, spread the table cloth wet it with clean water, and cover it over with a spare cloth, by placing a few bamboos across the top of the table, and throwing the spare cloth over all. The boiling being over, and all ready to pour off the contents on the table, turn the cock, or take out the pin at the bottom of the boiler, as may be, and let the indigo run on the table. As soon as this is done, a part of the fecula will find its way through the table-cloth to the bottom of the table, and, from thence, to the drain at the end of the table. Take this up in *chatties*, or if a pump is fixed to the end of the table, pump it up into the table through a strainer, until the water appears of a dark-red tint, free from colouring matter, and then let it run off through the saving vat, if one is in the factory, or otherwise away altogether.

Ninth.—In five or six hours, if the table has a wide spread, which it should always have at the bottom, the water will have run off, leaving the indigo covering the bottom of the cloth. Then scrape up the whole of it to the corner of the table, place a weight upon the top of it, and leave it a few hours to thoroughly cool and drain itself; during which time have the presses in readiness and the cloths spread inside of them and wetted ; after which, take up the indigo, and fill the press boxes. Lay on the cover of the box first, and, after a while, drop the beam upon it. Soon after, commence screwing gradually, until, in about five hours, you have got the indigo down from 8½ inches to about 3 inches, and no water oozes through the holes at the sides of the boxes ; then take off the pressure very evenly and gently, by moving both the screw-nuts above the beam at once and equally, and open the sides of the press boxes for cutting the mass into cakes. This part of the operation requires great care and attention, from the apathy of the natives, who too frequently neglect turning the screws, and then, to make up for lost time, turn away as fast as they can. By doing this the cakes are sure to be cracked and spoiled. Take care of this, go on slowly and carefully, and all will be well ; otherwise, assuredly not.

Tenth.—The indigo being pressed into a square mass, mark it with the end of the cutting knife, with a 3-inch flat rule, or as may be, wider or narrower according to the designed size of the cakes ; in the middle of which stamp the factory mark. Then divide it, either with the knife, or by a brass wire, (whichever your cake-cutter is most expert at,) in squares, and let them be taken to the drying-house, and placed on the *machans ;* each day's manufacture being placed separately, with a ticket to distinguish the same. The best plan to prevent mistakes, or roguery amongst the native servants, is to stamp the cakes with the number of the day they were made, beginning with No. 1 the first day, and continuing on the numbers to the end of the season.

Eleventh.—The indigo being on the stages, keep boys to turn them until dry,—the cake-cutter and rung-mistrie also should be in attendance,—and take especial care that no draught of wind runs through the drying-house, otherwise much breakage may be apprehended.

Twelveth.—The indigo being considered all but dry, have the whole placed in the sweating room ; and with the cakes build up a consolidated wall ; distinguish each day's work by placing a few shreds of *sunn* between them ; the whole of the indigo being disposed of in this way, and the solid wall formed, cover it over with blankets and dry *bhoosah* (bran), and shut the door, securing the outsides, so as to prevent the possibility of any wind entering into it. In about 15 days the indigo will be ready to go again to the *machans*, and should be removed thereto, keeping each day's work separately as before. Proceed now cautiously, and gradually remove the outside coverings of the door, and then the *bhoosah*, &c., from off the heap of cakes, a little by little each day, for four or five days; it being a grand point not to expose the indigo too soon to the effects of the external air, as sudden exposure would crack and break the cakes to pieces.

N. B.—The above process is advantageous, since it gives a brilliancy to the colour of the indigo, and the now-so-much-prized white skin to the cakes.

All factories do not have sweating houses.

Lastly.—The indigo being thoroughly dry and fit for packing, and your chests (generally made at the factory of mangoe-wood tarred) ready, clean the cakes, but do not polish them; then begin packing; placing equal quantities of indigo in each box. Should you not have sufficient of one day's work to fill up with, note the quantity of each day's cakes—so many of No. 5 and so many of No. 27 (as may be) in chest No. — and be sure to let the muster cake truly show the mean quality of the indigo in the chest it represents. Stow the chests as lightly as possible, keep each layer of the cakes horizontally even with their marks upwards ; nail the covers of the box close down to the indigo, that in moving the chests about afterwards, it may not be broken by having play inside ; with the chests send the muster box, and, thus, the whole process ends.

———

CHAPTER IX.

INDIGO IN BENGAL.

IN the Lower Bengal indigo factories, whose lands are yearly flooded over by the rise of the river Ganges, the system of sowings is entirely different to the Tirhoot mode. No drills whatever are used in Lower Bengal, either for October or February sowings; the seed in all factories is sown broadcast, either in wet or dry sowings. The wet sowings are done as follow: On the subsidence of the overflow of the river Ganges, about the 1st of October, men are sent out along the banks of the rivers, and particularly to the *churs* or islands formed in the river's course, and wherever a likely deposit of thick, soft mud has formed (alluvial deposit), the seed is there scattered broadcast by the sowers, each of whom is provided with a long bamboo to help him through the mud — which is often waist and occasionally neck-deep—and a bag, or, more frequently, a cane-work basket of indigo seed, to scatter. These baskets are replenished from time to time by the *amun* or head-factory servant, in attendance in a small boat called a *panchorey* or *dinghy* (see illustration); this servant is responsible for the sowings being

XVII.

(Cheetanee) Sowing in Lower Bengal.

Thacker, Spink, & Co., Calcutta.

XVIII

Lower Bengal Planter's Boat (Bhowlea).

properly done, and must also see that no seed is wilfully wasted or destroyed by being thrown into the water, or on to deposit which is only *sand* un-mixed or with but little mud, which, of course, will never give a crop.

If the deposits of the year from the overflow of the Ganges are found to be very deep, the sowers will sometimes employ a raft of three plantain-tree stems fastened together, which they will pole over the mud and slushy places, carrying the seed basket with them. These seed baskets are woven of jungle cane, stripped smooth and wound round and round in a spiral form, and tightened together by strips of cane ; they are then quite water-tight, and will float like a wooden bowl, even when holding a considerable quantity of indigo or other seed, and they are very durable and useful, being handy for all manner of purposes.

The Assistant will, during these *cheetanee* sowings (from the Bengali word *cheet*, signifying to scatter), go about mostly in his *bhowlea*, or house-boat, an illustration of one of which is given. These boats are manned by four or six rowers and a steersman or *manjee ;* they also have one or more large square lug-sails, and some are fitted up, schooner fashion, with jibs and fore and aft sails ; but as keels are most dangerous to have on a boat, more especially a sailing boat, on the Ganges, on account of the swift stream and rapid eddies, these fore-and-aft rigged vessels are dangerous to life, and a mistake. By a kind provision of Providence, also, and the eternal fitness of things, keels are unnecessary on the Ganges, for the downward current of the river will carry the planter, or any other traveller by water, down stream at a good rate, while for progressing up stream the southern breeze blows, with hardly an exception, every day during that portion of the year in which the Ganges is navigable by large boats ; so that as the stream runs from north-west to south-east, and the wind blows steadily from nearly south-east to north-west in the opposite direction up stream, there is no need for sailing in a wind such as makes a fore-and-aft rig necessary, and the square lug is found to be the best and most powerful sail, easily dropped in a squall, and soon up again. Besides, the mud banks and islands in the Ganges would make travelling in a keeled boat most dangerous, as, if the keel struck on a mud bank with the stream running one way, and the wind blowing in the direct opposite, a capsize would be unavoidable ; whereas, with a flat-bottomed boat with a good beam, no danger is to be apprehended, as, the sail being promptly lowered, she is easily " slewed " off (see illustration).

When the indigo seed is thrown on the alluvial deposit, it sinks from two to three inches by its own weight, and, in the course of a week or so, the plant springs forth. *Cheetanee* sowings are easily and cheaply done, so far as the sowing of the seed itself only is concerned, but they are most expensive to weed, as with the indigo plant spring up countless varieties of jungle plants, weeds, and creepers, and specially a sort of jungle called *jhow*, something like a small kind of fir or pine, which is very hard to eradicate from the ground ; in fact, it is next door to impossible to get quite rid of it,—relay after relay springs up as one lot is cut and piled into heaps,—in never-ending succession the plants appear, and they are a source of infinite trouble and expense to the planter to get rid of. When the hot, dry months come round in April and May, these *churs* or alluvial deposits crack in every direction, and are most dangerous to ride or hunt over, as, a horse getting his foot into one of them will fall to a certainty, or throw his rider, perhaps severely. The best pace to go over them, if they must be traversed, is a quick-march step, a sharp walk, like a cat on hot tiles—and the feelings of the cat in that predicament and the rider across a very " cracky " *chur* must be much alike, so far as we may compare human feelings of uncertainty and alarm with those of the lower animals.

The indigo plant springs up in this new soil with great vigour and freshness at first, but as the fiery heat of April and May tell upon the plant, it remains stunted, and withers up to a mere little stick, as wiry and thin as a piece of a knitting-needle, and with two or three shabby little leaves on the top as an apology for verdure. But they are very hardy, however, and when the rains commence, about July, these little plants shoot forth at the first heavy shower, and, as the rainwater penetrates to their tap-roots through the cracks in the top soil, the plants spring up with renewed vigour, and soon throw out a fine show of handsome leaf.

Cultivated sowings, performed much after the Tirhoot style, are also sown in Lower Bengal, in October, besides the *cheetanee* sowings. The ploughs used are the same, but the seed is thrown broadcast, and no drills for sowing are used ; neither are rollers used to break the clods, nor hand-labour, but the clods are pulverized by dragging a couple of large bamboo frames—resembling ladders, and called in Bengali *mooies*—over the land until the clods are sufficiently reduced in size. If this operation is delayed until the freshly ploughed-up clods dry and harden in the sun, it is impossible to obtain a satisfactory result. After every ploughing, the *mooies* should, therefore, be promptly applied. When the seed is sown, the *mooie* is applied for the last time, and the land left for the plant to spring, as in Tirhoot.

The process of weeding, cutting, carting, and manufacturing the indigo plant are almost precisely the same in both Tirhoot and Lower Bengal, except that, in the latter part, boats are very largely used instead of carts to transport the indigo plant to the vats of the factory, on account of so much waterway being available owing to the many streams of Lower Bengal, and also to the fact that the main portion of many factories' sowings are close by the river-banks, either of the Ganges or its tributaries.

When the manufacturing season is drawing near, a very sharp look-out has to be kept on the rising of the Ganges, as, often, a great deal of plant sown alongside the river is swamped and lost in a sudden and unexpected rise of the water; therefore it is advisable, in manufacturing to get all plant in danger of inundation near the rivers cut away first, in order to be on the safe side. The boats generally take the plant right up to, or close under, the vats by a canal made for the purpose, if no natural waterway exists; so that it is undisturbed and unsoiled from the time of its being cut and placed in the boat till it is placed in the steeping vat—which is a great advantage, as in carting plant along muddy roads for long distances much of it is unavoidably soiled and rendered

almost unfit for manufacture with the finer and
cleaner plant ; the cart-wheels, scrubbing against
it, tear away the plant and plaster with mud what
is left at the sides.   The Assistant will find travelling about in his *bhowlea* not at all unpleasant, as
these boats are comfortably fitted up (like the sketch
given) and contain every accommodation; in fact, they
are very pleasant, except for the heat during the
middle of the day.   But as this is the time when
everyone stops work for a while, the boat can be
anchored, or tied up to some convenient tree, or in
some shady nook, or by the side of a village on the
bank, and the three or four most sultry hours of the
day passed in comparative comfort.   There is always
a breeze on the water, and the boat will be pretty
quiet, as the men will, generally, at this time, be on
shore cooking their food for the mid-day meal.   In
fact, the Assistant must learn to look upon his boat
as his moving home for the time being, and will, of
course, take stores, tea, sugar, etc., and his cook and
khitmatgar about with him.   During the sowings the
bhowlea is to the Lower Bengal Assistant what the
bamboo-cart is to his *confrère* of Tirhoot.   Although
the sowings in Lower Bengal would, in many instances,
horrify a Behar planter, yet for quality and colour the
Lower Bengal indigo will always be better than the
up-country made indigo in the North-West Provinces.

CHAPTER X.

THE subject of indigo in the North-West Provinces has not, it seems, been sufficiently understood. The network of factories in the North-West are mainly native, in some few instances managed by Europeans—absolutely European proprietorship is scarce. Under the native system, where a European manager is entertained, the practice is generally as follows : We premise that from 300 maunds of plant, 1 maund of dye is extracted. In the month of January advances are made by the purchasers of the plant, seed, or *gâd* (unpressed indigo) to the *zemindars* or *ryots*, but more generally to the former, for security's sake. The intending purchaser ascertains the growing capabilities of the land which he intends to sow with indigo, and the number of ploughs the contractor can command. Assured on this point, he enters into a calculation by which he satisfies himself that 10 *katcha bighas* of land are the maximum which one contractor can manage to cultivate properly with the aid of a single pair of oxen. The produce from this will average from fifteen to twenty maunds of plant and one maund of seed per *bigha*.

R., In.   F

The *katcha bigha* is an unknown quantity in Lower Bengal,—a little more than ⅓ of a Shahabad *bigha* of 5½ *haths*,—2½ *katcha bighas* go to one *pucca bigha* of 5½ *haths*, or, roughly, ⅝th of an acre.

An agreement is formally drawn up, in presence of witnesses, by the *gomashta* (or head servant) of the factory, and to this document the usually illiterate contractor affixes his mark, or he may simply touch the pen with which the contract has been drawn up. In a few instances, he is able to sign his name. The deed sets forth the extent of land to be cultivated ; the quantity of plant, *gâd*, or seed to be delivered by the contractor ; and the rate to be adhered to, which, for the plant, is, generally, from Rs. 4 to Rs. 6 per hundred maunds, and for *gâd*, Rs. 12 per maund.

One hundred maunds of plants will, as a rule, yield the planter 2½ maunds of *gâd*. It is stipulated that it be delivered in such a consistency that 3½ seers, or about 6 ℔, can be taken up with the hand. It is also customary to pay the contractor in advance the full value of the plant or *gâd*, which he has guaranteed to deliver, though another system is also in force. In the latter instance the terms are as follow : one half in advance, a quarter after the first irrigation has been completed, and the remaining portion after the first weeding. Indigo fields are generally irrigated before the setting in of the rains, and they are weeded twice before the plant arrives at maturity.

If the contract be concluded with the *zemindar* instead of with the *ryot*, as is generally the case, a higher rate of about Rs. 22 per 100 maunds is paid by the purchaser to the contractor. If the contract be made with a *ryot*, he brings forward the *zemindar* of his village as his security. The same strict supervision is exercised over the *zemindar* as over the *ryot*, and the same forms are gone through in both cases. Having received his advance in January, and paid a time-honoured one per cent to the *gomashta*, the contractor does what he pleases with the money ; but he must sow between April and June.

Each factory has a number of peons called *sepahis*. These servants look after the welfare of their master, keep the contractor to his engagement, see that the *ryots* irrigate the indigo fields and otherwise exercise a strict supervision over *zemindar* and *ryot* alike. Without such unflinching surveillance, the engagement would run but a poor chance of being properly fulfilled, and the purchaser would be the inevitable loser. In the event of idleness on the part of the contractor, he is brought up to the factory and admonished, or he may receive warning through the peons. These men do not get *dasturi*, or hush-money ; but it is feared that they too often batten in other ways upon the poor contractor when they visit his indigo fields.

When a *ryot*, who has contracted with the purchaser, does not come up to time with his rent to the *zemindar* or proprietor of the soil, the standing crop of indigo is liable to be attached for rent. Where this happens, the purchaser always pays the rent due and realises it from the *ryot* afterwards. The *zemindar* seldom gives annoyance in a matter of this kind—he is amply satisfied with having obtained his rent. The payment of rent by the purchaser is never one of the conditions of the agreement, nor is such an occurrence by any means frequent.

The indigo harvest commences in August, when a portion of the plant is reserved for seed, which ripens in November. The factory-accounts are made up in October. A *ryot* contracting to supply indigo is supposed to realise a profit of one rupee per *katcha bigha*. He has, however, other sources of income. He has paid his year's rent for his indigo fields, and he makes the best use of the land after the indigo is cut ; whilst, in the event of the *khuntis* (second cuttings) not being left to produce both fecula and seed the following year, the cultivators prepare the land for a *rabi* or cold weather crop which comes to maturity in January. The *khuntis* are, in their turn, dug up, and melons sown, at the roots of which, before the vegetable is quite ripe, indigo seed is again cast. Thus the cultivator obtains three crops from the indigo field during the year.

That the indigo cultivation is the least remunerative of the various kinds of agricultural speculation is certain. The reason why, it is followed, is, that the large advance in January is a temptation not to be resisted : it is so much ready-money in the ryot's grasp ; he is a rich man for the hour, and he is happy. Moreover, he has debts to meet, and a demand for rent which will shortly be made. In some parts prosperous *ryots* even cultivate indigo without advances.

How indigo has flourished under the liberal Government of the North-West Provinces will be seen by the following figures :—

	1849-50.	1858-59.	1880-81.	1881-82.
Doab		16,202	45,434	57,042
Benares	cannot	14,896	14,484	15,710
Bengal	find.	49,225	17,425	18,957
Behar		26,420	58,062	58,569

In this table " Benares " includes Gorruckpore, Patna, Shahabad, Ghazeepore, Azimghur, Benares, Tounpore and Allahabad. " Doab " includes all places above Allahabad, the distinction being made on account of the difference in the quality of Indigo produced.

More striking difference.

	1843-44.	1857-58.	1877-78.	1888 89.
Doab	6,400	9,360	44,285	64,000
Benares	16,400	10,000	17,556	18,600
Bengal	97,000	50,330	16,502	17,200
Behar	23,400	18,822	34,857	58,748

Indigo cultivation may be divided into two great heads : *niz* and *assamiwar*. The former cultivation may be likened, in some respect, to a Home farm managed by the proprietor of an estate in England. It is prosecuted on lands of which the party or "concern" has acquired the tenant-right, or the right of actual occupancy, by an establishment of ploughs, oxen and servants maintained at the planter's own expense. Occasionally ploughs and oxen are hired for the purpose when the establishment kept up at the factory may not be sufficient.

The *ryoti* cultivation, on the other hand, as the very name implies, is carried on by the tenants on their own lands under contract or advances made by the planters as will be explained hereafter. But *ryoti* itself is divided into cultivation of two kinds : one carried on in villages or estates in which the planter has temporarily acquired the rights of the *talukdar* or *zemindar*—the other, in villages belonging to outside parties. These two systems are familiarly known as *ilaka* or *be-ilaka ryoti* cultivation ; the latter also frequently called *khuki* indigo. These same divisions have been in existence for over twenty years, and they hold good at the present day, except, perhaps, that the *be-ilaka* or *khaski* indigo has largely increased as compared with *Ilaka*, *assamiwar*, and *zirath*. *Khaski* has also largely extended in some of the Behar districts. In 1860 the amount paid for *ryoti* indigo in Tirhoot was Rs. 6-8 per bigha for good and Rs. 3 for bad or indifferent crops. At the same time, in Bengal Proper, according to the statements of some of the planters of that time, the rate for indigo was from 8 to 4 bundles per rupee,—a bundle generally averaging somewhat over two maunds in weight. We read in the Indigo Commisson Report, 1860, that Babu P. K. Tagore, in answer to the question, " How many bundles did you take for one rupee ? " replied : " Generally 8 ; but I reduced it to 6, which amount was not remunerative to the *ryot.*" In the Jessore and Hooghly Districts, 8 bundles are frequently mentioned in the Report as being received for 1 rupee, whilst in Jessore, Babu C. P. Chaudhri, in his evidence before the Commission, stated that he received 10 bundles in the Nuddea District. The *ryot* was always charged for seed. In paragraph 148 of the same Report, the witness states that the produce in bundles from a bigha was 12 to 24, and the average 16 bundles. The price of indigo during the time when Babu C. P. Chaudhri carried on his business, averaged Rs. 250 per factory maund. The advances generally made were Rs. 2 per bigha. Looking back at the prices paid for indigo in these days, it will be seen that the *assami* has now fallen on much better times, when he receives his advance of Rs. 18, 20, or 22 for his *khaski* indigo. It might be noticed that in Behar the *khaski assami* generally receives his seed free and suffers little or no vexatious supervision from the planter.

The only time he is required to be present, is when the European Manager or Assistant measures his fields. This is generally done twice—once to settle the advance, and, again, after the crop is up, to certify that the proper amount of land has been sown and to see what the crop is like. In some parts the former Bengal bundle-system has been improved upon by the introduction of the maund-system-rates varying from 7 maunds (equal to about 3½ bundles) to four maunds (equal to about 2 bundles) to the rupee being paid in different parts of Behar. The average rate is, however, about 5 maunds to the rupee, an amount which admits of a very handsome profit to the *assami*.

The question of indigo-cum-opium has frequently attracted attention—occasional alarm being caused by the opium agents clashing with managers and tenants alike. There is absolutely no reason for any such friction, as opium and indigo are grown at different seasons, and there is nothing to hinder the *assami* growing his opium from October, and his indigo from March, making a paying speculation out of both. Sugarcane, tobacco, and potatoes are now the chief rivals of opium, and the first named crop has, of late years, made serious inroads on the opium cultivation of Shahabad. Were the Sóne canals as reliable for cane irrigation as they ought to be, the opium cultivation in that district would be considerably more reduced than it is. The price paid for opium has risen, but not so high as the times would warrant, and the profits made from the monopoly by Government keep suggesting, still fairer and more reasonable treatment of the cultivators.

CHAPTER XI.

THE crush of indigo factories, both European and Native, in Behar, has induced me to attempt, in this chapter, a descriptive outline of the nature of the various forms of tenures prevailing in the North-West Provinces. In doing so, I would ask the Assistant entering indigo to bear in mind that my account will have *special* reference to the district of Gorakpur,* which I emphatically believe to be *the* direction in which the industry of the "blue dye," forced out of Behar, will, in the near future, compulsorily seek its legitimate development.

I propose, therefore, to notice briefly—

1. The administration of the Rent Laws.
2. Jurisdiction of Revenue Courts.
3. Nature of tenures, and rights of tenants.
4. Kinds of soil, and average rates of each.
5. Settlement.
6. Liabilities of tenants.
7. Rights of Proprietors and *Ticcadars* in tenant leaseholdings.
8. *Sir* land.
9. *Mokarraries.*

---

* Gorackpore and Shahabad, in North-West Provinces in South Behare are in commercial circles reckoned as Benares District (indigo).

1. *Administration of the Rent Laws.*—All suits between landlord and tenant of whatever nature, in so far as they relate to land, must be filed before, and disposed of by, the Collector, or Assistant Collector. Appeal from either of which courts lies to the Commissioner of the Division, and, ultimately, to the Board of Revenue for the North-West Provinces, whose finding is final. We are sorry to have to record that such suits are of too frequent occurrence, and that their name is legion. We have suits for arrears of rent, enhancement, ejectment, distress, determination of rent, and nature of tenure, assistance to eject, and many others filed by the *zemindar*, or his *locum tenens pro tem.*, against the tenant ; for excess of rent collected, or for fixing the proper rental of his holding, compensation, wrongful dispossession, abatement of rent, etc., the mere enumeration of which our limited space forbids.

2. *Jurisdiction of Revenue Courts.*—These courts have a very extended jurisdiction, and in all suits under the Rent Act, tried and decided by the Collector of the District, or by the Assistant Collector, their judgment is final, except in cases where the value of the matter at issue exceeds Rs. 100, when appeal lies to the District Judge, or where the amount of the suit exceeds Rs. 5,000, in which case the appeal lies to the High Court. In all other

cases, appeal from the order of the Collector of a District, or an Assistant Collector of the first class, lies to the Commissioner of a Division, whilst, in that from an Assistant Collector of the second class, the decision of the Collector is final.

3. *Nature of tenures and rights of tenants.*—In the North-West Provinces the words "tenant" and "tenure" have a wide and many-sided significance. It might casually be presumed to mean, respectively, the *ryot*, or cultivator, and the holding of lands for which he pays rent ; but this assumption would be wrong, and would in no way embrace all the constructions put upon the term by the Rent Act (XII of 1881). By the provisions of this Act, the word "tenant," not only includes the cultivators who inhabit the villages and cultivate the lands pertaining to them, but the *kadars*, 'sub-lessees,' and *katkanadars:* 'sub-sub-lessees' come also within the meaning of the term. With these latter, however, we have little to do in this chapter.

Tenancy divides itself into four great heads, *viz.* :—

    (*a*) Occupancy tenants.

    (*b*) Non-occupancy tenants.

    (*c*) Tenants at fixed rates.

    (*d*) Ex-proprietary tenants.

In order to render clear the several natures of the above holdings or tenures, we must briefly refer to the system prevailing in the North-West Provinces—the so-called *lumberwari* system. By this system—and here the North-West Provinces have enormously the advantage of Bengal every field (cultivated, grass-land, or otherwise) in the village or *mouza,* even including cattle tracks and the village site, are num-bered at the settlement, and the names of the then tenants of each number recorded in the *khesarah,* whilst the number is again clearly inscribed on each plot in the *naksha* or map of the *mauza.* Thus, each tenant can, by reference to the *patwari* (or village-accountant), obtáin from him a signed record of the numbers of the fields in his possession. The number-ing of each field secures the tenant against the shifting about or exchange of lands, and the confu-sion consequent upon this which obtains, with such miserable results, in Bengal.

A.—Occupancy tenants.—Every tenant who has held unbroken possession of the same lands under cultivation for twelve years, has obtained a right of occupancy in that land. The occupation, or culti-vation of the father, or what person soever the tenant may legally have inherited from, gives the tenant in actual possession at the time the right of occupancy. But a right of occupancy cannot arise from a gift of land to the tenant by the landholder. The right must, in all cases, be acquired in strict

accordance with the provisions of the Rent Act.   Nor
is the occupancy right transferable, although the
actual cultivating occupation of the land for a time
as proprietor, and afterwards continuously as a
tenant for twelve years, confers such right.   No right
of occupancy can accrue in *sir* land (*vide* notice
page 95), in land held in lieu of wages, in land
occupied under a written lease, or in waste land held
for grazing purposes.   The phrase " actual cultivating
occupation " has been held by the Board of Revenue
to mean that, actual, and not vicarious, which occupation creates, by the lapse of a term of twelve
years, a just title to occupancy rights.   But it has
been held that such rights can be acquired in the
following instances.   When a tenant occupies or
cultivates alluvial land, which may or may not be
capable of continuous cultivation, for twelve years ;
when lands in which occupancy rights have already
accrued have been submerged, and again exposed
after a lapse of years, the failure to pay rents during
the period of submersion does not debar if the newly
formed land be identical in site ; when lands are
held under a written lease, but the lease not having
been registered is invalid, and cannot bar the tenant
from acquiring occupancy rights, even although he
may have used the lease as valid.   This last exception to the stringent provisions of the Act is worthy

of the careful notice of the planter. When a *tikadar*, or lessee of a village, claims a right of occupancy in lands included in his lease, and can show that he has actually cultivated them himself continuously for twelve years.

B.—Non-occupancy tenants are tenants who have not occupied and cultivated the same fields for the required period of twelve years and who can be deprived of their tenures at the will of the land-owner, on being served by the latter with a written notice of ejectment under the provisions of the Act. This notice is compulsory, and it is under no circumstances left to the option of the landowner. Such tenants are also called tenants-at-will. Their rights are few, and their chances of retention, after legal process by the landowner, scant. A non-occupancy tenant may dispute his liability to be ejected, by application to the Collector or Assistant Collector, such application being made within thirty days after service of notice. Tenants of this description, having no fixed status, may be finally dealt with in this paragraph, and further remarks will not be found regarding their *liabilities* in touching upon that subject.

C.—Tenants at fixed rates.—This class is only found in permanently settled districts, and the term " at fixed rates " signifies any tenant who has held

land—'either personally or by heritage from his predecessors'—in interest, from the time of the Permanent Settlement, at the same rates of rent. The right
of such tenant is transferable, and may descend by
succession ; but his rent cannot be enhanced, except
on the ground of increase of holding by alluvial
deposits or otherwise.

D.—Ex-proprietary tenants are a class with whom
the indigo planter can have little to do ; and we cannot do better than quote the Act, section 7 : " Every
person who may hereafter lose or part with his
proprietary rights, in any *mahal*, shall have a right of
occupancy in the land held by him or *sir* in such
mahal at the date of such loss or parting at a rent
which shall be four annas in the rupee less than the
prevailing rate payable by tenants-at-will for land
at similar quality, and with similar advantages."
" Persons having such rights of occupancy shall be
called *ex-proprietary tenants,* and shall have all the
rights of occupancy tenants." The rate of rent of
this class of tenants is fixed by order of the Settlement Officer, or by an order under the Act, and the
tenant may apply for abatement of his rent, either
on account of diminution of the area of his holding
from diluvion or other causes, or on account of the
decrease in productive power of the land from any
circumstance over which he has no control.

In the present chapter we have but touched upon the more salient points connected with the nature of tenures and the rights of tenants, whilst our space naturally renders our remarks crude and incomplete. To the young Assistant who would dive more deeply into the sea of the Rent Law, we would recommend a little book published by the Government Press, Allahabad, " The North-West Provinces, Rent Law Manual," by Mr. L. W. Teyen of Fatehgarh.

4. *Kinds of soil, and average rates of each.*—Under this heading we would ask our readers to bear in mind that in this portion we refer chiefly, indeed entirely, to the district of Gorakhpur. The *bhât* land is pre-eminently suitable for the cultivation of indigo, and the few factories we know are built in, and cultivate, lands of this description exclusively. A few ventures have been made in *bangar*, but with ruinous results. The peculiarity of *bhât* land is its retention of moisture, which obviates the necessity, to planter and *ryot* alike, of irrigation. Indeed, it is only in years when the rains have entirely failed, that we have seen the tenant of the *bhât* soil resort to irrigation, either before or after the sowing of his crop. Thus, the nature of the soil gives him the advantage over his, comparatively, less favoured brother of the *bangar*. I am not agricultural chemist enough to give the analytical reasons for this difference in the moisture-

retaining properties of the two soils. *Bhât* land is generally subdivided into five minor classes of soil—*kat-bhât* (bastard *bhât*), *dhoos* (sand), *soombha* (rich loam), *bal-soombha* (sandy loam), and *matiar* (clay). Of these the *soombha* and *matiar* are in most favour with tenants, as affording the qualities of soil which best adapt themselves to the producing of *rabi* (cold weather) and *bhadai* (rainy weather) crops, respectively. Planters, on the other hand, do not much affect *kat-bhât* or *matiar*—a small proportion of the latter kind being, however, desirable, as being the most moisture-preserving soil in dry seasons. *Soombha, bal-soombha*, and *dhoos* are the classes of land after which the heart of the planter yearns, and in which the indigo plant loves to take deep root and flourish mightily. The rates of *bhât* land vary a good deal from various causes, which I am not called upon here to explain, but I think I shall not be far out in stating an average all round rate at Rs. 2-8 to Rs. 2-12 per *bigha*. Time was when this rate would have been considered enormous, even within own recollection ; but the rates of land have been and are, steadily on the increase, as a hungry population, forced out of Behar, pours in.

Bangar is a soil which, of itself non-retentive of moisture, invariably demands irrigation for the success of its crops, although after the rains the tenants are sometimes able to sow without previous irrigation. Why *bangar* lands should not be able to keep their moisture, even in years when *bhát* lands from excessive rains are regularly saturated, is a matter hardly within the province of this chapter. It must suffice that the fact is indisputable. There appears to be no doubt that *bangar* is the richer, stronger soil of the two, yielding, perhaps, on an average, 15 per cent. more produce than the *bhát*. The rates are pretty much the same in both soils. In the district of Gorakhpur, *bangar* soil is of three kinds—*safed*, or white; *lal*, or red; and *karaili*, or black *bangar*. These three qualities seem equally productive, the first-named kind being predominant in Gorakhpur.

5. *Settlements.*—In the North-West Provinces the settlement is permanent or terminable after a period of thirty years. The Gorakhpur District is at present under settlement, which will probably be finally concluded within the next four years. Contrary to form erusage, the merits and capabilities of the soil will be the standard of assessment, and this would seem a sounder and more practical plan than that of assessing according to the prevailing average rates of rent paid by the greater number of tenants in each *zemindari*.

6. *Liabilities of tenants.*—All classes of tenants are liable to be sued for arrears of rent, which rent

still remaining due, he becomes further liable for interest at 10 per cent. per annum.   If, however, the arrears still continue unpaid by the 30th of June following the date of decree, the tenant shall be ejected on application made by the landowner.   Notice is issued, and, failing payment within 15 days from receipt of notice, the tenant will be ejected.

All tenures, of whatsoever class, are also liable to be enhanced, under conditions peculiar to each, for the nature and process of which enhancement we would refer the reader to the Act.

The produce of all land is liable to be distrained for arrears, being hypothecated for rent due in respect of such land.   It is also optional for the landowner to substitute recovery of rent by distress and sale of the produce for the more tedious process of a regular suit.   Standing crops of every kind, and crops which have been reaped and gathered into the *khalian* (threshing-floor), are liable to be distrained provided only that they are the produce of the identical lands for which arrears of rent have accrued.

The tenant is also liable to have a suit filed against him, for the determination of his rent and the nature of his tenure.

7. *Rights of Proprietors and Tikadars in tenantless holdings.*—The proprietor or *tikadar* has a full right to dispose of any land from which the tenant has

been ejected, either by appropriation or by re-arrange-
ment with a fresh tenant. When the landowner or
sub-lessee thus appropriates for his own cultivation
such vacant holding, he becomes liable for the rent
fixed, and the land thus acquired is denominated *sīr*.

8. *Sīr lands.*—The *ziraths* of an indigo concern
come under the classification of *sīr ;* and the proprietor of the concern becomes the tenant of the land
so held, with all the liabilities, but none of the rights
of the ordinary tenant, within the meaning of the
Rent Act. The right of occupancy would not seem
to accrue in *sīr* or *zirath* land, for a *tikadar* or
lessee of a village must prove that he has *himself
actually cultivated* the land in which he claims the
right continuously for twelve years. In the case of
Hardeo Baksh v. *Kallee*, Mr. Reid, the Senior
Member of the Board, observed : " It is simply pre-
posterous to decree right of occupancy to four men
in 323 *bighas* of land which they are shown to have
held as lessees. The cultivation or occupation, which
is the basis of acquisition of a right of occupancy,
is the cultivation or occupation of a *cultivator*, not of
a *lessee.*

9. *Mukarraries.*—This term is applied to lands or
villages which are leased for ever at a fixed rate of
rent, which cannot be altered by the landowner except at the settlement, and then only when the lessee

refuses to pay any enhanced rate at which the lands or villages may be assessed by the Settlement Officer. *Mukarraries* are simple leases in perpetuity, at a fixed annual rental. No class of tenant can lease his holding in this way. Indeed *mukarraries*, although common prior to the passing into law of the Rent Act now in force, were irregular and illegal, having only the sanction of the crass ignorance, as to the law of the landowner. The practice seems to have been borrowed from the districts of Saran and Champaran, in Behar, which border on the Gorakhpur District to the east and south.

I have now glanced at the principal features of tenures in the North-West Provinces, and would close this chapter by again referring the Assistant to the excellent manual previously mentioned for a more extended knowledge of this important question.

CHAPTER XII.

THE system of indigo cultivation adopted by European factories in the districts of Behar, south of the Ganges, is closely allied to that pursued by the almost universally *native* concerns in Benares and the North-West Provinces generally. The soil is of a similar kind, and red *bangar*, which, unlike the *bhât* soils of North Behar (the great seat of the indigo industry in India), is non-retentive of moisture. Consequently the sowings in Shahabad and contiguous districts are chiefly carried on during the rainy season—July, August, and September.

The earlier (rainy weather) sowings are called *assarhi*, and are cut in September-October of the same season. The later sowings have no special designation at sowing time, though in the following season, when the plant, having been pruned, is cut, it is then called *khunti*. Perhaps *nauda* would be a more appropriate name (one which has been applied to *assarhi* as well), and, in the following pages, *nauda* will mean the plant which is sown in August-September, and cut in the following season during July-August. *Assarhi* signifies plant sown in *Assarh*, the

Indian name for the month which corresponds to the last fifteen days of June and the first half of July. The *assarhi* crop is a particularly precarious one, and it is only in years of light and favourable rainfall that it yields a good return. Even then, the quantity and quality of the produce is vastly inferior to that of the *nauda*. *Assarhi* is never exclusively sown in all concerns in Shahabad and other districts for the above reason.

The stumps of the *assarhi*, after being cut in September-October, are left in the ground, and in favourable years, generally give another cutting along with the *nauda* (both at this stage being called indiscriminately *khunti*) in the following year ; but, frequently, in the heats of spring, the plant dies away, thus presenting another serious objection to *assarhi*.

The *nauda*, after germinating and becoming sufficiently advanced to allow of the process, is weeded carefully, and in December, January, and February, it is again cleaned and pruned. Sometimes a second cleaning and pruning in April-May is necessary, but not always.

Nauda may be looked upon as the backbone of the Shahabad indigo system, strengthened and bolstered up by *assarhi* as described above, and *jamana* or *assamiwar* indigo. Under this latter system, the *ryot* or tenant takes money advances from the factory on the understanding that he will sow so much indigo in opium lands, after the opium season is over. He is paid, according to agreement, either by the *bigha* or by weight ; the rates in some places, in my time, used to be six factory *maunds*, weighed on a steel-yard, to the rupee. At other factories, various rates per *bigha* prevailed, ranging from ten to twenty rupees. The crop under consideration was sown towards the end of March and all through April, after the land had been relieved of the previous opium crop ; irrigation being effected generally by means of a *mote* or *moth*. The following description of the apparatus is about the best I have seen, and I give it here in full :

" The *moth* consists of a leathern bucket made out of single cow's hide, varying in capacity from 1½ to 3½ cubic feet ; the edges of the leather of the bucket or bag are turned over an iron ring from 1¾ to 2 feet in circumference, and three iron stays welded to the ring at equidistant points at one extremity, and also welded to one another, provided the necessary attachment for the rope.

" This rope, generally made of buffalo's hide, 1 inch in diameter, is passed over a rude wooden pulley supported by a forked rest, firmly fixed at the lower end, and inclining over the well's mouth

at the upper end.   The rope is fastened to the yoke
of the pair of bullocks, which supply the motive power.
The bullocks descend an inclined plane varying in
length with the depth of the well, and thus bring the
bucket to the surface, where it is emptied by a man
specially stationed for the purpose.   In the field under
irrigation, a third man is employed in admitting the
water to the crops, or in sprinkling it over the ground
with a wooden scoop, shaped something like a spade
(called a *hatha*), from the spots where it collects.

" The average capacity of a *moth* is nearly 2¾
cubic feet, the quantity of water raised per hour
varies from 75 to 250 cubic feet, and, as the
working day is often not less than twelve hours, the
daily outturn varies from 900 to 3,000 cubic feet.

The value of the leather bucket varies from 3*s.* to
6*s.*, the rope costs 8*s.*, and the ironwork, 1*s.* 6*d.*
The bucket lasts during the season of irrigation (about
four months), and the rope two years.

" The cost of irrigation for one season, as estimated
by Mr. H. C. Levinge, the Superintending Engineer
of the Sone Circle, is as follows :—

		£	s.	d.
One-tenth of the cost of a pair of bullocks	...	0	7	3
Four months' feed of bullocks ...	...	1	4	0
Four months' wages for the labourers	...	2	0	0
Cost of *moth*	...	0	11	0
Total	...	4	2	3

The average area irrigated is eight acres to one moth ; so that moth - irrigation costs about 10s. per acre for the year."

CHARGES FOR IRRIGATION BY SONE CIRCLE OF CANALS.

Annual leases.

		Rs.	A.	P.	
Rice	...	... 3	0	0	per acre
Bhudai	...	... 2	8	0	,,
Rabbi	...	... 2	8	0	,,
Sugarcane	...	... 5	0	0	,,

Hot-weather crops for irrigation between 1st April and 25th June.

All crops ... Rs. 4 0 0 per acre.

Five-year leases.

All crops ... Rs. 2 0 0 per acre.

If water is taken for sugarcane in hot months the total charge is as follows :—

			Rs.	A.	P.
Under 5-year leases, per acre		...	2	0	0
During hot months	...	...	4	0	0
		Total	6	0	0

This is simply that the hot-weather rate is charged over and above this for 5-year leases.

The Government is not altogether favourable to the Indigo-cum-opium system, in Shahabad at least, and there are several factories there that have lost, of late years, a good portion of their *assamiwar* cultivation owing to the *ryots* getting the idea that it would please the Government officials if they refused to take advances. A native Deputy Magistrate (Uncovenanted Civil Service) told the author during the time Sir Ashley Eden was inquiring into the indigo systems of Behar, that the Lieutenant-Governor did

not want any more planters in the country ; following
up the above inscrutable remark by dismissing a case
of ours which was before him. Not many months
after we had another case before this wiseacre. This
time he fined one *jemadar* ten rupees for assaulting a
carter—oracularly vouchsafing the remark that, as the
Sahib had not come to court in person, the suit as against
the luckless *jemadar* must necessarily be true. I have
not infrequently experienced similar displays of like
behaviour from the small fry of magisterial magnates.
This same gentleman, in yet another case where
a villager was sued for deliberately upsetting his
bullock-load of indigo plant on the roadside, delivered
himself of a curiously " mixed " decision, sentencing
the man to 15 days' imprisonment. This finding
was afterwards quashed by the High Court of Calcutta, when they remarked that the Magistrate had
acted under no section of the Penal Code, but had
with more ingenuity than circumspection, appro-
priated one from the Civil Code instead. Our Deputy
had made a mess of the case, and got duly drawn
over the coals for it, though, in the other cases,
where the miscarriage of justice was greater, he was
not so much as admonished. The same gentleman,
when had up at Arrah as witness in some Registration Frauds, was found by the Judge to be unable
to decipher his own English decisions. These are

only a very few of the shortcomings of native magis-
trates whom Lord Ripon had contemplated making the
future arbiters of the English men and women in India !

Yet another description of indigo has sprung into
existence since the opening of the Sone canals,
which is called *falguni* from the fact of its being
sown in the month of *Falgun* (February-March)
In this kind of cultivation, *niz* or private lands,
i. e., lands in the actual rented possession of the factory itself as tenant and cultivator, are irrigated by
canal water and sown at the end of February and
on into March. This system has been found very
remunerative in the few European concerns in the
North-West Provinces, though the sowings are,
owing to the cold, somewhat later than in Shahabad,
which has a considerably warmer climate than the
North-West. The following schedule will enable the
reader to see at a glance the different kinds of crops,
their duration, &c :—

Crop.	When sown.	When reaped.	POSSIBLE NUMBER OF CUTTINGS.		
			First year.	Second year.	Third year.
Jamooali	March-April	July-August	One	*None*	*None*
Assarhi	June-July	Sept.-Oct.	One	Two	One
Nauda	Aug.-Sept.	July-August	*None*	Two	One

The reader will perhaps pardon a slight digression here, whilst we briefly describe the famous Sone canals.

The magnificent scheme of Colonel Dickens, by which Shahabad should be irrigated by a widely ramified system of canals, which canals should also be navigable, was presented to the Government of India in 1855, but, obstacle after obstacle presenting itself, the final sanction was not obtained until 1871. Preliminary operations had, however, been in hand since 1869, in which year the construction of the Dehri-on-Son Anicut was commenced. The main western canal having Dehri as its starting point, though partly dug out to Chunar in the North-West Provinces, remains still a long and expensive monument of Government folly and wastefulness. It has not been completed beyond Bedadee, two miles west of Sasseram, on the Grand Trunk Road. Except Bhabua and the Rohtas Pergunnah in Sasseram Subdivision, the other divisions of Shahabad have, by the construction of the Sone canal, acquired a complete immunity from famine.

Briefly stated, the following facts are available regarding the Sone canal in Shahabad. The main western canal has to carry 4,511 cubic feet of water per second to irrigate 1,200,000 acres, only 600,000 of which require irrigation at one and the same time.

The Arrah canal (which is navigable), with its branches the Beeheea and Doomraon canals, commands an area of 441,500 acres under *kharif* and *rabbi* crops.

The requirements of water are taken at one cubic foot per second for each 133 acres. The duty of the water for irrigating purposes is arrived at in this way. The irrigable square mile is considered to contain 250 acres, 140 being deducted for roads, tanks, village sites, &c., and the remaining 250 as not requiring simultaneous irrigation. As one acre of rice requires 27 cubic feet of water per hour or ·0075 cubic feet per second, a square mile of 250 acres requires 188 cubic feet per second. The canal itself has four principal distributaries, exclusive of the Beeheea and Doomraon canals. The Beeheea canal has seven distributaries, and the Doomraon canal twelve.

The Buxar and Chansa canals command an area of 309,500 acres under *kharif* and *rabbi* crops.

Bhabua, already referred to as likely to suffer in time of scarcity, has, including the thannahs of Ramgurh and Chand, an area of 1,037 square miles, containing 1,307 villages, and a population, according to the census of 1872, of 294,252, giving a population of 284 to the square mile. There are several rivers or large *nuddies* that flow through the Bhabua Subdivision which might be used for irrigation purposes. The Suri and the Kara have in other dry years been bunded up by the Chenpore and Khelaspore Zemindaries.

The first *mahai* or manufacture, called *morhun* in Tirhut, commences in Shahabad sometime in July, and consisting of the *nauda-khuntis* and *jamooali*, yields the best produce and colour. The second *mahai* continues till the end of October, and consists of *durzies* (local name for second cuttings of *khunti*) and *assarhi* plant. Towards the end of the *mahai*, produce is very inferior, whilst the indigo made is black and hard. Manufacture in Shahabad is, unlike Tirhut, mainly carried on without the aid of machinery. The necessary apparatus consist of from 6 to 15 pairs of vats of 800 to 1,200 cubic feet per vat, a boiling-house and press and cake-houses to contain the boilers and the implements necessary for pressing and preparing the cubes of indigo, a reservoir, the cubic capacity of which is at least sufficient to hold water enough to fill a quarter of the factory vats before complete displenishment. The reservoir is kept filled by *moths*, (a description of which appliance has been given), the water being taken from either natural or artificial tanks, a lake, a river or wells. A bungalow, offices, stables, and out-houses complete the *tout ensemble.* These are generally located in a strip of land taken under perpetual lease or *mukarrari*

This holding is usually limited to a few acres, though some old factories have acquired large estates in various ways, partly by buying, partly by perpetual leases, by acquired occupancy rights, *gozasta khastkari*, in fine, by all, and every means rendered legitimate by local use or legal acceptation.

As regards the lands on which the planter grows his indigo. These may have been acquired in any of the above-mentioned ways; but, if new lands have to be catered for, the procedure resorted to is multiform, and the offspring of circumstance or opportunity. In the factory which I managed for a number of years, the following system of acquiring lands was preferred :

The factory *jemadar* or bailiff would go round the village under negotiation, noting carefully those lands which he thought would be suitable for our purpose. I would then, after selection and approval, send him to the holder of the lands with a detailed application for the area which was required. Should the landholder agree, the fields were forthwith measured, and a lease written out, giving us possession from May of one year to August-September (full moon) of the next inclusive, or 15 months and some odd days. This seemingly unequal term was in order to admit of our cutting our plant in July-August. For, had the lease been drawn out on the same terms as usual,

R., In.   H

we should have had absolutely no return for our money.

The rent was all paid in advance, and I had no-thing more to do than cultivate and sow the lands in their proper seasons. I should mention here that there was always a conditionary clause in every lease, enabling us to extend our possession of the lands for another year at half-rates. But this term was entirely optional. In this way, if the *khuntis* were strong and healthy-looking, we could hold on to them for second cuttings, with the further prospect of a third year's crop.

This land arrangement was an arduous task, and it was made still more so by the fact that it had to be repeated every year. In seasons when grain was plenty, land was always scarce, and one had to take what one could get ; but when famine pinched and hunger pressed, the pick and choice of the best lands in the countryside was ours.

The concern where I spent some five years of his sojourn in Shahabad in comparative comfort con-sisted of four factories. My first billet was at Masai, afterwards Chainpur became his head - quarters. During my incumbency of these factories, my sole amusement was shooting ; and, as all the places lay at the foot of the Kaimur hills, I had plenty of it.

———

INDIGO IN MEERUT.

THE relations between cultivator and factory in these parts is somewhat different to that which obtains in Behar. Here we have indigo either by the advance system, *badni* or by *khaskreed*, that is, purchase in the open market of indigo plant, grown by the cultivators as a speculation. In Tirhoot and Behar generally the *assamiwar* or *ryothi* system to some extent prevails, but the *khaski* system now more generally obtains. The cultivators living in the villages within what is called the factory *dehât* take advances to sow indigo for the factory at a certain price. This price may be anything from twelve rupees to four annas per maund, and is a matter of mutual agreement. The seed is generally given to the cultivator. This is not the case in the North-West Provinces, there the cultivator generally grows his own seed. In Sarun the *khaski* system has become very popular, and the cultivators look forward to their indigo advances more eagerly than they do for the opium advances. The advances given are, in some instances, heavy, and the conditions on which the money is given are not, in most cases, half as harassing as those inserted in the *badni* agreements of the North-West Provinces. By the term *factory dehât* is meant the circle of villages within which a particular factory claims the monopoly of growing

indigo as a vested right from usage.   We might give
as a parallel instance the right of shooting or hunting.   A man who shoots on preserves is called a
poacher, one who builds a factory and grows indigo
in another factory *dehât* is called an interloper.   This
right is not so much exercised in the North-West
Provinces, where, consequently, free trade in indigo
prevails.

That interesting work, " Garden and Field Crops,"
published by the Agricultural Department, North-
Western Provinces, says :   " The advance or *badni*
system was probably brought about by the *assamis*
requiring some guarantee that the indigo grown by
them would be purchased by the factory according
to the *badni* system.   In March and April when the
crop is sown, the factory binds itself to purchase plant
at rates then agreed on ; the rates are fixed considerably lower than they are in free competition ; but as
long as sixteen or eighteen rupees per 100 maunds
of plant is paid, the system is not more objectionable
than that followed by Government in furthering the
opium cultivation.   Unfortunately one of the principal objects of the factory in making advances is
often not so much to arrange for a crop in the present as to gain a power over the cultivator as will
enable it to compel him to grow indigo on its own
terms for the future.   The power once acquired may

be used to compel him to grow plant at the factory's will, and to sell at a price much lower than it would otherwise command."

The following memo. of the cost of an acre of indigo in Meerut is also from " Field and Garden Crops " :—

		Rs.	As.
Ploughing, clod breaking, sowings, &c., &c., say,		1	8
Price of seed, say,	... ...	1	8
Weeding (this is often not done at all)	...	3	0
Reaping	... ...	1	9
Rent on *kharif*	... ...	2	8
Watering	... ...	4	15
	Total ...	15	0

As a return for this expenditure in a year like the present, the following may be taken as pretty near the mark :—

75 maunds plant @ Rs. 14	Rs.	9	8
2 ,, seed @ ,, 6 per maund	,,	12	0
Total ...		21	8

CHAPTER XIII.

INDIGO IN THE AZAMGARH DISTRICT.

THE manufacture of indigo for export dates from the early years of British rule in the district.

The Company's commercial resident was allowed to trade on his own account, as well as for his masters, and Mr. Crommelin, commercial resident at Azamgarh and Mau, in company with two gentlemen. named Stewart and Scott, started the first indigo concern in Azamgarh. Its establishment was opposed by the magistrate of the district under the rule which forbade the Europeans to occupy land, and engage in indigo manufacture in the ceded provinces without permission from the Governor-General in Council. Mr. Crommelin eventually got leave to hold the factories, and was exonerated from blame in connection with them ; but so aggrieved did he feel by the action of the magistrate, that he filed an action in the Supreme Court for damages. What the result of the action was, the records that are within reach do not show. His indigo concern seems to have been conducted through European and Eurasian assistants, and at first, not without disputes with the natives, which the latter were careful to bring to the notice of the magistrate. The prohibitory rule under which Mr. Crommelin had been opposed, did not apply to persons born in India, and soon after the establishment of his concern, others were set agoing in various places. In 1808, Mr. D. O. Ferguson, acting on behalf of a Major Stevenson, erected a factory at Nizamabad, and another, now fallen completely to ruin, was established at Imilia, which, though professedly the property of a native, was, in 1811, managed by a European named O'Dell. As time went on, more factories were erected at various places. In 1812, Mr. Ferguson's factories passed into the hands of Mr. J. Sturmes. This gentleman carried on the concern vigorously, and added to it another branch factory. He also engaged with sugar and cloth trades, and, by mortgage, purchase, and farm, held possession of a number of landed estates. After his death in 1821, his indigo concern and estates were managed for a number of years by his executors. In 1829, the indigo concern and part of the estates were sold—the former being purchased by Mr. H. E. Hunter. Meanwhile Mr. Crommelin's factory had changed hands ; some being held by persons resident in the district, others by persons resident in Calcutta or elsewhere, who managed them through agents. For several years after 1829, Mr. Hunter, who, in addition to the Nizamabad concern, had taken over some of the

factories erected by Mr. Crommelin and others, and
Mr. J. H. Stonehouse who held the Dobrighat concern,
were the chief independent Europeans in the district.
They traded largely in sugar and other native produce in addition to indigo, and Mr. Hunter was also
in possession of a good deal of landed property
He died at Nizamabad in 1845, and his large indigo
concern was broken up. Mr. Stonehouse also had
been unfortunate in business, and had to relinquish
most of his factories, but he continued to reside in
the district, holding a small factory, which he built at
Rajapatti in Pergunnah Nizamabad, till 1857. At the
time of the mutinies there were at least nine concerns, whose headquarters were in the district. The
chief of them was the Dobrighat concern held by
Mr. E. F. Venables. Previous to the mutinies, and
for some years after them, the production of indigo
remained chiefly in European and Eurasian hands.
The only native who seems to have held factories
before the mutinies was Basu Darzi, who, having
been originally Mr. Hunter's tailor, and afterwards
his chief manager, had acquired a good deal of
money, and on Mr. Hunter's death, purchased six of
the factories. About 12 or 15 years ago natives
began to take an interest in the trade, and the comparatively high prices of 1862 and the following
years brought about among them a rage for factory

building. Numbers of new native factories sprung up. Of 415 factories now standing in the district, 323 have been built during the last 14 years. At the present time only 29 factories with 115 pairs of vats are the property of, or held by, Europeans and Eurasians. The rest (386 factories with 607 pairs of vats) are the property of, or held by, natives.

CHAPTER XIV.

INTERLOPERS AND INTERLOPING.

IT is not very many decades since the above words were unknown in India, or if known at all, they had no application to Indigo. At the opening of the present century, indigo—the *Devil's Dye* of the credulous of those days—was still in its infancy, and the pioneer of its progress had to be imported from the West Indies. Doubtless even then the few and scattered servants of " John Company " may have looked askance at the intruder, who, the protecting ægis of letters patent notwithstanding, had the pluck and hardihood to entertain and prosecute the five years' scheme—a tentative one—of the magnates of Leadenhall. Jealous enough of adventurers of all sorts refusing permission to settle in India unless in rare exceptions except under pledge and bond—not to adhere to the strict letter of which was disaster— deporting from India at a moment's notice, and all property and goods confiscated by the Company, the Hon'ble Company was not always sufficiently alive to its position to welcome and foster an element in its population which came adorned with the triple armour of English pluck, capital, and intrepidity.

The first principle of the old established factory, the foundationstone upon which the " man in possession " builds his exclusively protective claim, is nothing more nor less than that of *vis major*, the proprietor of an indigo concern, having nestled down comfortably in any tract of country, satisfied himself that he has thereby acquired an inalienable lien upon the whole tract to the exclusion of all others. Argument carries no conviction with him—the interloper is his *chien du diable* beyond the pale of mercy, beneath all consideration, to be hounded out and ruthlessly " smashed " wherever encountered. Such arguments as these find favour with the pseudo-monopolist : " I am first in the field. I have, at great cost, and with much diplomatic difficulty, established friendly relations with the *zemindars*, conciliated the trading community or *mahajuns*, who, seeing in my arrival on the scene a saviour of the needy landholder, their immemorial victim, opposed me with the myriad wiles known to their usurious cult, but who quickly came to see that my object was, not ultimate possession, but terminable tenure. I have been a blessing to the peasantry, to whom my coffers were always open ! I have rendered to the Government of the country an inestimable service in the better farming of the soil on my own part ; and in the beneficial effect of my example upon the

ryots who are in immediate contact with me. To
the extent of my leases, I have secured to the revenue
of the country the invariable payment of its demands;
to the tenant I have been a fostering father, and have
rescued him from a tyranny not to be endured. I have,
in short, nurtured the growth of numerous blessings in
the tract in which I have settled, and some consideration should be shown me for these and other reasons."

The arguments of the interloper are in many res-
pects similar to those of his opponent. The interloper pledges himself to do as much for the tract
of land he cultivates and its population, as may have
been accomplished by the earliest settler.

Some of the prominent causes which have led up
to this new source of friction amongst the members
of the Indigo body corporate, are as follow :—

1. *Crowding out, the result of oversupply.*—Of late
years the numbers of boys sent out from Home to
" try their luck in indigo " have increased beyond all
reasonable limits. The Home market is glutted, and
it behoves parents of a certain social status to seek
employment for their sons abroad. They turn their
eyes upon our great Indian empire and find that
even in that vast country the well-born young man
must choose his career absolutely either in indigo,
tea, or an office in one of the Presidency-towns.
This young person may or may not have been

plucked for either the Army or the Indian Civil Service. Thus, with a mind which had been buoyed up
with a false hope, until the time of the fatal examination all his thoughts are still of India. His family
knows, or may be related to, some one who has been
engaged in planting. Influence is made, and the re-
jection of the examination, or the neophite of the
public schools, is shipped finally for India. The
strong purposeful boy soon forges ahead, and wins
the vacant management ; but others less fortunate
swelter and stew on £12 to £15 a month all the days
of their natural manhood ; and when they become
old, they die or disappear. So, the fittest survive
not always ; it is, also, too often, the survival of the
luckiest. The hapless ones despair, die, disappear,
run into debt, or become interlopers. And how is the
metamorphosis achieved ? Thus :

2. The liberal-minded and educated native *zemindar* is rapidly on the increase. Such men, shrewd
and appreciative of commercial gain, as are all natives of the country, have begun to see the enormous
profits accruing to the proprietors of indigo factories,
who are their *tikadars* or lessees, and the matter
suggests itself in so practical a form, that they
are slow to project operations on their own behalf
Leases fall in and are not renewed, even under the
most alluring conditions ; the *zemindar* has " struck

oil," and will build him a factory. The old preju-
dicial influences, inculcated by caste restrictions, or
ghostly fear, have disappeared ; the *zemindar* is
wealthy and will be his own banker, to whom
no interest accrues ; the ziraths of our capitalist
over-way, stretch tenantless around him ; and he has
but to cast about him for the malcontent Assistant
who is able, and but too willing, to help him.

    3.   *The small outside capitalist or would-be planter*
—a recalcitrant and aggressive person, who has clear-
ed his *lakh* or so in various contracts, " wise and otherwise," and who sees in indigo a hopeful venture in
which to invest his " pile."   Under the eyes of the
very protectors who had, in many instances, furnished
him with the means to fortune ready made, this
unruly reprobate would " rear his crest " and interlope.
What indigo man but calls to mind the flatulent
fortitude of such men as the quasi - truculent conspi-
rators—the hitherto automatic dumb waiters on their
masters, the planters,—whose mighty minds would
henceforward be devoted to the construction of pre-
datory vats.   The outside capitalist and his tribe buy
shares in old established concerns, and, failing in
ambition to outshine the sun, are feign to twinkle
lowly little stars afar off, and to comfort themselves
with the thought that were but the great orb away
each twinkling star would be a sun.

4. It often happens that successful young Assistants club together and, with no immediate prospect of promotion where they are, amalgamate their slender savings and hie them forth to build. With unassuming modesty they refrain from the rude shock of unequal contest with their seniors, and betake themselves to tracts where indigo is comparatively unknown, or where only one or two pioneers have preceded them. Here land is cheap and a perpetual lease easily obtained. True, the soil is churlish, and yields but niggard crops even to the frugal *ryot.* What then ? It shall be improved, they think, with better treatment and a liberal application of *seet* or indigo refuse. It is a wide unclaimed country this, to which they have got. Have they not been obedient sons of mother Association,—cheerfully refraining from any act which could render them arraignable as interlopers? The nearest working factory is good 26 miles away : surely no interference is to be expected or warranted here They will take a modest little perpetual lease on this high spot in the chief village on this opportune little river which never dries up they are told, and does not call for any Saran canal schemes to ensure a supply of water for *mahai.* Perhaps it may be a little feverish ; but then, they are pioneers, and endurance is the first article of the pioneer's faith. They will

do very well, little by little, and, thank Heaven! there is heaps of room to extend. What a lovely country it is, to be sure, save an occasional official on tour, the place seems really *terra incognita* to the white man.

Slowly, however, the little cloud, not nearly so large as a man's hand, appears just about the horizon, and it spreads out towards the young enterprise and becomes threatening. What is this little cloud which threatens to envelope, to overwhelm the new empire? Whence the strange flutter in the dovecots of the speculators? Good 26 miles away is the factory of Mr. Green. He has heard of the "interlopers," and if money can do it, he will oust them. The simple *zemindar*, who had willingly granted a lease—who had known no European save our young friends—to whom the small advance had been salvation, who had been proud of the confidence placed in him by the *sahibs*, and the high position of friend and patron, the *zemindar*, has been worked upon by still more glittering offers, and unstable as water, he sells to Mr. Green. Right and title, the power to give or hold, option to grant or smash the quite conditional lease which the adventurers had taken, in all good faith, from the *zemindar*, all are vested in Mr. Green. But he will be merciful. He too had struggled. "*Topie, topie ek*" will be amply vindicated in their future easy dealings with Mr. Green. Ah too trusting ones, Behold your ultimatum! Move on, or be expelled! Choose, and quickly!

Rebellion is in the hearts of the interlopers. Heaven help them! what may they do? The Calcutta House who had so frankly taken them up and volunteered to tide them over difficulties by a mere mortgage of block and crop—the Agents themselves submit to Mr. Green and withdraw. Mr. Green is inexorable. They curse him and move on. They do not despair, for they advance still further into the jungles; and, late in the season, they build two vats, drop money and a partner, and toss rudderless on a sea of trouble across which it is doubtful whether they will ever pass.

It will thus be seen that the question of interloping is at the present moment a most urgent one. Indeed, rules have been framed by the Indigo Planters' Association which are designed to discourage interloping as between Europeans themselves. The folly of attempted intrusion, where space for development is insufficient for the man already in possession, is too self-evident to call for further notice here. But reasonable men have never been able to see why jealousy should intervene to debar men from building and pushing a business in tracts or *dehats* where there is room enough for all.

R., In.   I

CHAPTER XV.

IT appears that the cultivation of indigo, in the form practised in the West Indies, was commenced in Bengal between the years 1770 and 1780. Mr. Lowis Bonnaurd, who arrived in Bengal in 1772, and who had previously made experiments in Indigo at the Mauritius, is believed to be the person who is entitled to the credit of having introduced the manufacture of indigo to India. He did not succeed in the Mauritius in consequence of the soil not proving favourable to the growth of the plant. He made his first attempt of indigo manufacture in the French territory, about two miles to the southward of Chandernagore, in a garden called Gandalpara, with a single pair of small vats, the remains of which were to be seen a few years ago. The same spot is now occupied by a very excellent but small, factory, recently constructed. In 1775, Mr. Bonnaurd built a small factory at Faldanga, a little village between Chandernagore and Chinsurah. This factory was demolished some years since, and its site converted into a garden by its present proprietor, a native. About the above period, two French medical gentlemen,

Messieurs Ferrier and Michelet, supported by an enterprising English gentleman, erected a factory at Sampooker, near Ghiretty, which is still in existence.

After the East India Company became proprietors of a great extent of country, and the West India British subjects had abandoned the cultivation of indigo, the East India Company, acting upon principles similar to those upon which they had formerly declined the importation of that article, thought it expedient to encourage the cultivation of it in their own territories, for the purpose of promoting the industry of their own subjects, and of insuring, independently of foreigners, a regular supply of an article so essentially necessary to the most important of the British manufactures.   With this view, about the year 1779, they entered into a contract with an enterprising individual in Calcutta, upon such terms as induced him to engage in the cultivation of it, and his example was soon followed by others.   The Company continued to foster this revived branch of trade at a considerable loss, which, in the course of a few years, amounted to £80,000.

When the cultivation appeared to be fairly established, the Company resigned the trade to the gentlemen employed in their civil service, the free merchants residing in India under their protection.   The speculators became involved in distress, and the East India

Company thought it necessary and proper again to support the industry, by advancing money on the security of the produce to the amount of nearly a million sterling. This enabled the growing indigo industry to go on successfully, while it tempted many to engage in it, the result of which was a great extension of factories.

The marked advance of the indigo trade achieved by the exertions of European cultivators, has been justly adduced as a proof of the superior energy of individual industry ; the early fostering care and substantial support of the East India Company should not, however, be forgotten.

Previous to 1783, several factories were established, and a considerable quantity of indigo produced, and from that period it progressively increased. In 1783 the export from Bengal to Great Britain was only about from 1,200 to 1,300 maunds. At the same time, the importation of indigo into England from other countries, was upwards of ten times as much as that quantity. The Court of Directors in the letter, dated 22nd April 1789, to the Bengal Government, say : " We are in hopes, the measure of laying open this (indigo) trade will be attended with the good effects expected to result therefrom, and that hereafter it may become a permanent and advantageous article of commercial remittance, as well to the benefit of Bengal as of this country. In order to effect every possible improvement in this article, we transmit you herewith copy of a letter from William Fowkener, Esq., Secretary with Lords Committee of the Privy Council for Trade, covering a report of some experiments that have been made by a manufacturer of this country, with some hints necessary to be attended to in the management and preparation of the same. As it is probable, the information therein contained may be useful to the gentlemen concerned in indigo plantations, we direct that the same be made known in a manner that shall be most likely for rendering them publicly useful."

In the report above alluded to, Mr. Taylor (of Manchester, February, 1789) observes: " I suppose the East India Indigo [samples sent from Bengal] marked *A*, to be worth eleven shillings per pound, and equal to any Spanish Flora or Guatimala ; that marked *B*, is only worth five shillings and sixpence *C*, eight shillings and sixpence, and *D*, six shillings and ninepence, per pound ; and adds, that the result of all these experiments confirms him in the opinion, that the East Indies can furnish every kind of indigo wanted in Great Britain."

CHAPTER XVI.

THE following minute by Lord Macaulay, dated 1837, may be worth preserving :

" To say that the ryots of this country are mere children, and ought to be specially protected, is, I conceive, quite incorrect. They are not intellectually inferior to the peasantry of other countries.

" I am sorry to find that opinions differ widely on the questions submitted by Government, and that they differ most widely on the most important of those questions. That great evils exist, that great injustice is frequently committed, that many ryots have been brought, partly by the operation of the law, and partly by acts committed in defiance of the law, into a state not very far removed from that of partial slavery, is, I fear, too certain. But I see no reason to believe that any of the measures respecting which the Government has consulted planters would, in any material degree, alleviate these evils. Some of these measures, indeed, are quite unexceptionable, and would, as far as they go, operate beneficially. I would certainly give to the Sudder Amins jurisdiction in civil cases in which Europeans or Americans might be concerned. The only objection to this which has occurred to me is that at present an Englishman has an appeal to the Sudder Supreme Court, in every case in which a native would have an appeal to the Sudder ; natives have an appeal to the Sudder in cases originally tried before the Zillah Judge. All causes in which Europeans are concerned in the Mofussil are now tried before the Zillah Judge. The Englishman, therefore, has a direct appeal to the Supreme Court. If the Government should give to the Sudder Amins jurisdiction over cases in which Englishmen are parties, our countrymen will be deprived of the right of appeal which they now possess ; and possibly some discontent might by this change be excited among them. But I do not conceive this discontent would be deep or extensive, particularly if the Government would, in the exercise of its undoubted power, appoint a few intelligent Englishmen to the place of Sudder Amins in those districts which contain a considerable number of European inhabitants.

" The regulation which gave to the indigo planters who had made advances to a ryot a lien on the indigo crop, seems to me highly objectionable in principle. But I do not conceive that, by rescinding it, the Governor-General in Council would give any sensible relief to that class of the population whose interests appear to be peculiarly the object of his solicitude. The question appears to be a question between the planter and the *zemindar*. It is not easy to see how it can be of any consequence to the ryot, which of the two may distrain on his crop. I have no reason to believe that the *zemindars* exercise their power with more justice or humanity than the planters. The *zemindar* has great reason to complain of the existing regulation. It transfers to others that undoubted right of distress which he formerly possessed. Two other people, by an agreement between themselves to which he is no party, are allowed to deprive him of what was his due, and in return the law gives him a remedy which is certainly less expeditious and simple than that which he anciently had. Exactly the same extent to which this regulation is a benefit to the indigo planter, it must be considered as robbery of the *zemindar*. The misfortune is, that when once Government falls into such an error, there is great difficulty in returning to the right path. In the very act of destroying old rights, we create new rights, which must be destroyed in their turn if we revert to the old order of things. The *zemindar* had great reason to complain when his lien was transferred to the planter. The planter, who has invested his capital in the indigo business on the faith of the regulations of 1823, would have some reason to complain if that lien were now taken from him and given back to the *zemindar*.

On the whole, I should be inclined to recommend that
no planter should be entitled, except, of course, by
virtue of a special contract, to which the *zemindar*
should be a party, to distrain for any future advances.

"With respect to advances already made, I would
leave him the remedy which he now possesses. I
must again repeat, however, that the question is one
in which the *ryots* appear to be very little interested.
The law of pounding is in an exceedingly unsettled
state all over India. And this want of certainty,
doubtless, leads to oppression and injustice every-
where. These evils do not, however, appear to be
peculiar to the indigo districts. At a proper time
it will be necessary for the Law Commission to take
up the whole subject. It is a subject on which it
will be impossible to legislate usefully without an
extensive inquiry into existing rights and customs.

"The plan of rendering invalid all contracts for the
delivery of indigo which are not registered, seems to
me highly objectionable ; it would be either useless,
or in the highest degree vexatious. If the present
mode of registration should continue, the proceeding
would be a mere mockery. An agent employed by
the planter would attend, and would register without inquiry, all the agreements of the planter with
all the *ryots*. There would be no examination ;
nobody would ask whether the peasant had made the

contract freely, or under the fear of personal violence, whether he had made it with a full knowledge of what he was doing, or under the influence of deception. If, on the other hand, the registration is to be made really efficient, if the registering officer is to exercise over the labourer the same species of guardianship which, in England, the Judges of the Court of Common Pleas exercise in certain cases of contract over married women, if the parties are to appear, if the ryot is to be privately interrogated, assured of protection, encouraged to accuse the capitalist, the business would be absolutely interminable. In some districts 30,000 contracts would probably require registration in a year ; these contracts, I believe, are generally made at one season of the year. A great number of registrars would be necessary to conduct the examination into all these agreements. And the registrar, entrusted with the conduct of such an examination, must be no common man. He must be not only a man of sense, but what in this country it is hard to find a man of independency and integrity, a man who will dare to stand up for a poor native against a rich Englishman ; it would be hard to find such functionaries in sufficient numbers.

" It would be absolutely necessary to pay them well. The charge would be immense ; and, after all, it may be well doubted whether the advantages which the labourer would derive from such a system of guardianship would compensate for the journey, the attendance, the trouble, and the loss of time.

" There are contracts which it is very desirable to register ; contracts which are seldom made, which are made for long terms, which are great events in a man's life, which are likely to affect the rights of third parties, and which are of such a nature that they cannot be made without considerable trouble and great formalities. The purchase of an estate is an instance. The security, which registration gives to all the parties concerned in the transaction, is highly valuable. The trouble of procuring registration is but a trifle in comparison with the trouble which must attend such an event. But to require that all contracts for all terms between all the capitalists and all the working people of great provinces should be registered, and registered with such safeguards as to make it certain that every contract is freely and deliberately made, would be to dissolve the whole frame of society. The general rule which is followed all over the world is this, that no judicial verification of a contract should take place till it is alleged that the contract has been broken. At present it is probable that not one contract in a thousand is, in any country on earth, the subject of a law-suit. If the immense majority of contracts were not performed without legal investigation and decision, the world could not go on for a day.   The effect of a system of registration, such as that which I have been considering, would be that every contract without exception would be the subject of legal investigation and decision before it was made."

THE PATWARI.

THOUGH the *patwari's* pay is apparently small, yet it will be observed that, in the villages of which they are the poorly paid accountants, they and their relations hold the best *jotes*. Government they ignore, the *zemindar* they laugh at. In their *bastas* (bundles of account papers) they keep the wonderful evidence of their frauds which they hide by forgeries and cooked accounts.

The *patwari* being by law the man appointed especially to keep ryots' accounts, to give them receipts for rents paid, and to account to the *zemindar* for rents collected, is the most important of the *zemindar's* servants. His *basta* (bundle of accounts) the village record of rates and rights, and he is the backbone of the Ryotwari system. At present, the *zemindar* who pays the *patwari*, has no control over him.

The *patwari* difficulty might be arranged in the present day of high English education as a connecting link between the masses and the Government. For the *patwari* to become the true friend of the *ryot*, he must first be entirely separated from the *zemindar*. I would propose that the *patwari* commence as a probationer in charge of one village on Rs. 6 a month, rising to Rs. 12; after his probationary period expire, he would be promoted to the appointment of *patwari* on Rs. 20 a month, with charge of, say, 15 villages. The next rise would be that of Assistant Superintendent of *patwaris*, the charge being a circle of 150 villages, and the pay anything from Rs. 30 to 50 per mensem. I would suggest an Assistant Superintendent of *patwaris* in each subdivision, and over all a Superintendent of *patwaris* on pay rising from Rs. 60 to 100 in the head-quarters of the district, where he would have the office and records, and be under the direct supervision of the Collector. This would enable work to be carried out systematically, and secure a better class of men, as well as do away with the hereditary monopoly of *loot* possessed by the present class of *patwaris*. It would secure the Government a large number of loyal servants, and provide for a still larger class of educated natives who at present have nothing to do. The money to pay this could be raised by a *patwari* cess to be realized from landlords only.

Though the usual fare of the cultivator in the hands of the *patwari* is the spider and fly business, yet when the tenant has the good fortune to be situated where the power for weal or woe has been rescued from *patwari* and *zemindar* alike, and is in the hands of the European middleman, the *Behar Indigo Planter*, the *patwari's* accounts which have done for the cultivator assume a certain degree of regularity, and tolerable order is evolved out of chaos. The honest, truth-loving, masterful Englishman has done, by his immediate and continuous contact with the *ryot*, what empiric and intangible law could never do ; he has frightened the *patwari* into order. This has, however, not been done without worry and headwork, for it is only one prepared to martyrize himself in unravelling the sinuosities of the *basta* (of the *patwari's* accounts) who can hope to do much.

In different localities the difficulty experienced in mastering the *patwari's* accounts differ in degree. There are places where it was quite possible for the manager or assistant of a factory to have his village and land accounts literally at his finger's end, and that without mental strain or bodily exertion. Various reasons may be assigned for this facility, such as provincial and local custom, &c.; but the most prevalent cause will be found in the system adopted, improved, and handed down by former managers.

If the *patwari* of Behar is compared with his brother official of the Eastern districts of the North-West Provinces, the comparison is not favourable to the latter. In Bengal, the *patwariship* may, or may not, be hereditary. In the North-West Provinces, the office goes by strict entail—failing heirs direct to the nearest of kin.

In Behar, the *patwari* is *nominally* the Government village account-ant, *really* the servant of the *zemindar*, by whom he is paid and dismissed at will ; in the North-West Provinces, he is quite independent of anybody whilst living miles away ; the local administration cannot exercise any efficient control over him. He occupies the best *jote* or plot of ground in the village or villages of which he is *patwari*. Owing to the system obtaining in the North-West of numbering each field and all cultivable or non-cultivable land in every village (such numbers being shown upon the *nuksha* or map, the names of the tenants being also entered opposite the numbers in the *khesrah* or index) the *patwari*, who is, *per se*, not entitled by law to acquire land in his official capacity, finds the names of brothers, wives, cousins come handy to serve his purpose, though that purpose may be in direct contravention of the statute. He is frequently the village *tahsildar* (rent-collector) as well. To enable him to fill this office, he keeps a double system of papers—one for the *Sarkar*, and the other, for the joint mystification of landlord and tenant. One he calls the *Sarkari jama-bandi*, in which rates are low and collections recorded as well up; the other goes by the name of the *tahsil-dehi*, or *gaoni-tahsil*, a bloated and ever wonderful record of wrongs. The Bengal *patwari* resides upon his *halka* or circle; in the North-West this gentleman loves to dwell in the bosom of his family, far away from the scene of action. There the oppression is vigorously prosecuted by his lieutenant, the *gomashta*, whose commission is received upon the sum total of extortion—an iniquitous percentage.

In Regulation XII of 1817, the Bengal Government attempted to deal with the *patwari* question, but the attempt has proved unsuccessful, chiefly because the *patwari* is like a regiment with no officers between the colonel and the private.

In section (3) the Regulation states that every village shall have a separate *patwari*, but, if necessary, the Board shall authorize one *patwari* to look over several villages, or that more than one *patwari* shall be appointed to one village.

DUTIES OF PATWARI.

(16). The duties of the *patwari* shall be—

First.—To keep such registers and accounts relating to the village or villages to which he is appointed, in such manner and form as has heretofore been the custom, or in such other mode as may be hereafter prescribed by the Board of Revenue, together with such further registers and accounts as may be directed by those authorities respectively.

Second.—To prepare and deliver to the *kanungo* of the pargana, at the expiration of every six months, a complete copy of the aforesaid accounts showing distinctly the produce of the *kharif* and *rabbi* harvest.

Third.—To perform all other duties and services which it has been customary for him to execute.

(17). The Board of Revenue will determine on the mode in which the accounts rendered by the *patwari* to the *kanungo* shall be brought forward by the latter and recorded in the office of the Collectors.

(18). The *patwari* is to be paid hereafter in the same mode as he is now paid, whether in money or in grain or in land or in any other legal manner whatsoever ; but it shall be the duty of the several Collectors to complete an account of the mode in which such payment is made in the different parganas or other local divisions of their districts, and to submit the result of their researches to the Board of Revenue or other authority exercising the powers of that Board, and it shall be competent to the Board of Revenue or other authority aforesaid, with the sanction of the Governor-General in Council, to increase or reduce the amount of remuneration paid to the *patwaris*, and to alter or modify the mode of its payment in any case in which sufficient cause for the adoption of such a measure shall exist.

(19). Whether no *patwari* has hitherto been appointed, the amount of the remuneration to the *patwari* who may be appointed under this regulation and the mode of its payment shall be regulated by the Collector with reference to the usage of the adjoining villages.

(20). If the remuneration which a *patwari* has heretofore regularly received, or which may be assigned to him by the Collector or other competent revenue authority be denied to him by the parties who have hitherto paid it or, who may have been directed to pay it by the said authority, he is at liberty to complain against the person so with-holding his dues to the Collector, who will proceed to an immediate investigation of the facts and decide according to the usage of the village, and the Collector is hereby authorized to compel payment of the amount due to the *patwari*, and to fine the offending party accord-ing to his situation and circumstances in life, provided always that the fine in no instance exceed fifty rupees.

(21). In all cases in which the decision of the Collector is to be governed by usage, it shall be made an invariable rule to insert in the original proceedings on the case the attested report of the *kanungoes* of the pargana as to the custom or usage in reference.

(22). Collectors of land revenue are hereby empowered to summon the *patwari* of any village or villages within their respective districts whenever there may be occasion for his attendance on any matter connected with the duties of his office and to require him to produce all accounts relating to the lands, produce, rents, collections and charges of the village or villages, the accounts of which may be kept by him and to examine him on oath to the truth of such accounts and on any other matters relating to such accounts or regarding the lands, produce, rents, collections, and charges of the village or villages to which the said *patwari* may belong.

When a Collector shall require the attendance of a *patwari* for the purpose above stated, he is to serve such *patwari* with a written notice under his official seal and signature stating the purpose for which his attendance is required, and the papers (if any) which he is to bring with him.

(23). If any *patwari* shall neglect or omit to produce his original accounts on the requisition of a Collector or to give his evidence respecting them, the Collector, is hereby authorized and empowered to cause the said *patwari* to be apprehended, and to order him to be confined in the dewani jail of the district until he produces his accounts, or shows sufficient cause for not producing them. In such cases, the *patwari* shall be sent by the Collector with a *rubkari* to the Judge of the city or *zillah*, stating the purport of the order passed against him, and the Judge shall, on those grounds, commit the *patwari* to jail, and detain him until he produces the accounts or until the Collector applies for his release.

(24). In like manner *patwaris* shall produce all accounts relating to the lands, produce, collections, and charges of the village or villages, the accounts of which may be kept by them respectively and furnish every information and explanation that may be required regarding them whenever they may be required by any Court of Justice in any suit that may be depending before the Court, and if any *patwari* shall neglect or omit to attend with his accounts when required for the adjustment of any matter or dispute depending in Court, the Courts are authorized to order such *patwari* to be committed to close custody until he produces the accounts or shows sufficient cause for not having produced them.

(25). In any case in which a Collector of land revenue shall have occasion to depute an officer to examine the accounts of any village or villages, he is authorized to require the *patwaris* to attend such officer, and the Collector is further empowered to grant to such officer a commission to swear the several *patwaris* whose accounts are to be inspected, inserting in the commission the name of each *patwari* to be sworn, and if any such *patwari* shall neglect or refuse to attend such officer with his accounts or to give his evidence respecting them, when duly required to do so by a written notice from the Collector, the Collector is hereby authorized and empowered to proceed against such *patwari* in the same mode as if he had refused or neglected to attend or to give his evidence before the Collector himself.

R., In.   K

(26). (Repealed by Act XII of 1876).

(27). In like manner any *patwari* who shall alter, fabricate, falsify, or mutilate the accounts of the village to which he belongs or shall furnish to the *kanungo* or Collector false, fabricated, or mutilated copies of those accounts, shall be held and considered guilty of forgery, and shall be liable, on conviction, to the penalties which are or may be prescribed for that offence, and any person who shall cause or procure any such forgery, shall be liable to the same penalties as those convicted of having actually committed the offence.

(33). In cases in which from local or other sufficient causes, it may appear impracticable or inexpedient to cause the appointment in any estate or farm of *patwaris* in the mode prescribed in this Regulation, as for instance, in certain estates consisting chiefly of hills and forests in the South-Western Frontier, and in very small *mahals*, the accounts of which are kept by the proprietors themselves, it shall be competent to the Board of Revenue to suspend its operation in such estates or farms.

Provided, however, that in all such cases the person by whom the village accounts are kept, whether proprietor or farmer or *gomasta* or other officer, shall furnish the *kanungo* of the *pargana* with such accounts and statements as the Collector, with the approval of the Board, may direct, and shall be subject to the provision contained in sections 22, 23, 24, 25, and 27 of this Regulation, and the proprietors or others by whom they may be employed, shall likewise be subject to the provisions contained in sections 26 and 27.

(34). No court of judicature shall take cognizance of the complaint of a *patwari* against the landholder or the tenants of a village for refusing to remunerate his labours, nor shall any court of judicature take cognizance of any complaint against a Collector for, or on account of, any decision passed by him in virtue of the powers with which he is vested by this Regulation.

(35). It shall be the duty of the Collectors to furnish the Board of Revenue with a periodical report of all judgments passed by them under section 20 of this Regulation, and such judgments shall be liable to reversal or alteration by the Board or Commissioner at any time within six months after passing the same, but not later.

(36). All sums adjudged by the Collector in favour of a *patwari* under section 20, and all fines directed to be levied by this Regulation, shall be recoverable by the same processes as arrears of public revenue, and all such fines, when recovered, shall be carried to the account of Government.

———

Extract from " Pioneer."

" This patient, underpaid, illiterate, tricky, and corrupt personage, is at the bottom of all the returns that form the recreation of Indian officials and the despair of statisticians at home. How fares it with the *patwari* in the wilds of the Central Provinces ? Mr. Fitzpatrick's report informs us that he is being civilised, educated, and otherwise brushed up in a variety of ways and with highly successful results. If we could see the *patwari's* side of the picture, we should probably find that he is having a very bad time of it in the Central Provinces. In the old days, he did pretty much as he liked, and no one asked any questions ;—If he was called on for returns, he sat in his house and evolved them ; if he was told to make a survey, he got some one else to do it for him ;—but these pleasant days are fast vanishing from the Narbada Valley, as they have vanished from these Provinces. The Land Revenue Settlements are falling in, the energetic Settlement Officer is abroad, and a Commissioner of Settlements, in the person of Mr. Fuller, has proclaimed the new doctrine that the survey and the record of rights must be done by the *patwaris* and not by a special and imported agency. The *patwari*, therefore, is made to attend school. He is taught mensuration and the use of the plane-table and chain. If he gets through this ordeal, he is kept hard at work through-out the cold weather, and often up to the rains, surveying his village : and in the rains he is given the inspiriting in-door employment of writing up and tabulating his cold-weather work. If this is not enough, he has arrears of current work in the shape of annual village returns and crop statistics to make up, and he is trotted out whenever the rains break for " field-inspections " and other pacific manœuvres. It is not surprising if some *patwaris* find the pace too great, and disappear from the ranks. The survivors form an army of trained and experi-enced land surveyors, who render the Province independent of the extraneous agency of a professional survey department. Compared with the cost of a professional survey, the work is being done by the *patwaris* and their overseers at an extraordinarily cheap rate, and with sufficient accuracy for all practical purposes. The Survey Officer objects to the comparison as unfair, on the ground that the Revenue

Officer gets his *patwaris* for nothing.   True ; but the *patwaris* must be
paid in any case, and from the point of view of the Local Government, it is more economical to get the work done for nothing by
existing agencies than to import a costly survey party from outside.
Nor is the further advantage of thus permanently improving the
efficiency of the *patwari* staff to be lost sight of."

*Extract from the judgment of J. Tweedie, Esq., Judge of Shahabad,
passed in Rent Appeals, Nos.* 510, 511, *and* 512 *of* 1886, *on the* 28*th
of February* 1887.   *In these appeals* " *nêg ;* " " *dak-beari* " ;
" *sarakh ;* " *and* " *saraf* " *were asked.* " *Batta* " *had been allowed
by the first Court, as being an equalizing item as between* " *Sicca* "
and " *Queen's* " *rupees.   There was no cross-appeal on this point.
The suits were under The Bengal Tenancy Act,* 1885.

       *   *   *   *   *   *   *

4.   These cases and all similar cases, must, in the absence of
' consolidation' *made with the ryot's consent,* be tried by the Courts
with reference to Section 74 of Act VIII of 1885.. That section is a
modern recast of the old law, and places the difficult question of what
are usually called vaguely, ' cesses,' in a clearer position than what
it previously occupied.   It must be prominently noticed, in the first
place, that that section by no means renders illegal *all* payments over
and above the " *asl-jumma.*"   *Only* " *impositions* " upon tenants are
rendered illegal.   An imposition may be taken to mean any payment
(in addition to actual rent) which is without a new ' consideration' to
support it ! And moreover, such 'imposition' must admit of being
appropriately called by the name of *abwab, mahtut,* or other " like
appellation."   Thus, if two different standards of money-measurement
are merely equalized ; here, the new state of things and the old are
identical, and there is not even an increment to the " actual rent " ;
far less, an "imposition" upon the tenants.   Or, if a *zemindar* supplies,
say, a sugar-mill for crushing sugar-cane (as is commonly used in these
parts) and charges for *its* use, when he charges for the use of his land ;
here also we have no "imposition" on the tenants, but we have a new
charge on a new " consideration."   So, also, if a *zemindar* engages
to keep all the village accounts for the tenants, providing necessary
establishment for the work, paper, pens, ink ; and so forth ; here also we
have—on a fair claim made—no " imposition " on the tenants in
addition to their actual rent ; but a demand for value received ;

a demand based on a " consideration " wholly distinct from, and in addition to, the " consideration" for which the " actual rent " is paid. Other increments will be recoverable under some law as the Road and Public Works Cess. Some extra payments, on the other hand, seem, by their very nature, to be " impositions upon tenants . . . in addition to actual " rent and to be fitly described by the words *abwab, mahtut* or other like appellation." Thus, when a *zamindar* tries to shift the burden of *dak-beari* from himself to a tenant, and demands payment, he is asking for payment without a " consideration ;" and is simply making an " imposition." Or again the same seems to be the case, when, under the name of " *Chunda,*" a *zamindar* seeks to recover from his tenantry, his own charitable or patriotic subscriptions. So, of other instances which might be given.

5. When, therefore, any plaintiff seeks to recover any items in excess of the *asl-jumma* as originally fixed, he is entitled to show and must show, one or more of the following facts, supporting h's case by appropriate evidence :—

I.—That the tenant had acquiesced in a new *asl-jumma*, in which items previously reckoned separately have been ' consolidated ' with the old *asl-jumma*, so as to form one new *jumma*.

II.—That some law in force gives him a right to the items claimed.

III—That the items are supported by a " consideration," distinct from, and additional to, the " consideration " in respect of which the *asl-jumma* " or actual rent " is paid.

IV.—That an ' extra item' is not really such ; but is merely an equalization of standards, not imposing any increase on the *asl-jumma*, or " actual rent."

6. As none of these facts have been established with reference to the extra items which are the main subject of these appeals, they were rightly disallowed by the lower Court. I have only to complete the subject by adding that all stipulations and reservations for the payment of " impositions," which admit of being appropriately called *abwab, mahtut*, or the like, are void. But this will not have any bearing or a stipulation to pay a new *asl-jumma*, nor will it require Courts to raise any question as to the calculations by which the new *asl-jumma* has been reached.

* * * * * * *

ARRAH ;   }   (Signed)   J. TWEEDIE,
The 28th February 1887. }   *Judge of Shahabad.*

Certified to be a true Extract.

Note.—Full Bench Ruling Chulban Mahton *v.* Tilakdhari Singh, Indian Law Report, Calcutta Series, Vol. XI (1885), page 175. *Held,* that *abwabs* are not recoverable, when it is not proved that they have been paid or payable before the time of the permanent settlement. *Patwaris' " nêg "* was included in the claim, but the Court made no attempt to distinguish it from the other *abwabs.*—A. P.

———

No. 381.

FROM

J. TWEEDIE, ESQ.,
  District Judge of Shahabad,

TO

THE COLLECTOR OF SHAHABAD.

DATED ARRAH, *the 15th June* 1886.

SIR,

I HAVE the honor to ask whether any order was ever promulgated for this District under sections 18, 19, and 20 of Regulation XII of 1817, relative to the remuneration of *patwaris.*

2. If any such order exists, please favour me with a copy of it, or a reference by which I may trace it out.

3. If no such order exists, I think that one is much needed, as a consideration of the Full Bengal Ruling cited below will show :—

Indian Law Report XI, Calcutta, page 175.

Patwaris' " nêg " was claimed along with numerous " cesses " and *abwabs.* All of these were disallowed as being illegal. No distinction was drawn in the judgment between *patwaris' " nêg "* and other " cesses ; " nor was the legality of *nêg* discussed with reference to Regulation XII of 1817—though attention was drawn to the Regulation in argument.

———

No. 2023G.

FROM

A. W. B. POWER, ESQ.,
  Collector of Shahabad,

TO

THE COMMISSIONER OF REVENUE, PATNA DIVISION,
BANKIPORE.

DATED ARRAH, *the 26th Nov.*, 1886.

SIR,

I HAVE the honor to forward herewith a copy of the letter addressed to me by the District Judge, alluded to in this office

No. 794G, dated 6th July 1886, and to draw your attention to the Full Bench Ruling cited therein.

2. In this District the mode of payment of *patwaris'* wages, which, according to the law, is regulated by custom (Sections 18, 19, and 20, Regulation XII of 1817) varies in different villages, in some the *patwari* collects his entire dues direct from the *ryots*, in others, the *ryots* pay the whole or a portion of his salary to the *zamindar* and he settles with the *patwari*. In villages where this latter custom prevails, the sums payable by *ryots* to their *zamindar* on this account, seem to me to be fully as legal a cess as the Road and Public Works Cess, both during their authority from express Statute law.

3. Some of the local Civil Courts following the Full Bench Ruling have disallowed *patwari's* " *nèg* " without inquiry into village customs, and the result has been in some places a wholesale repudiation of the obligation, I would, therefore, suggest that though the orders contained in Section 18, Regulation XII of 1817, have never been carried out, a beginning be now made first with those villages in which, owing to adverse decrees of the Civil Court complications have arisen, and also in those which are to be visited by a Deputy Magistrate (and Deputy Collector) in connection with the Amended Chowkidari Act.

4. The result of the Deputy Collector's inquiries in each village should, I think, be recorded in a concise proceeding which, when approved, would be sent in original for countersignature. A certified copy of this proceeding could then be filed with each plaint.

———

No. 1A.

FROM

H. J. S. COTTON, ESQ.,

Secretary to the Board of Revenue, L. P.,

To

THE SUPERINTENDENT AND REMEMBRANCER
OF LEGAL AFFAIRS.

DATED CALCUTTA, *the 13th Jany.*, 1887.

SIR,

I AM directed to forward the accompanying copy of a letter from the Commissioner of Patna, No. 1288R, dated 17th December 1886, together with its enclosure, from the Judge of Shahabad on the subject of the remuneration of *patwaris*, and to request that you will favour the Board with an expression of your opinion on the following points :

LAND REVENUE.

SETTLEMENT AND SURVEYS.

MR. REYNOLDS.

(*a*) Whether the provision of Sections 18, 19, and 20 of Regulation XII of 1817, authorize the Collector to take such steps as the Commissioner proposes ; (*b*) and whether if such a proceeding as is suggested were drawn up under the orders of the Collector, it would be accepted by the Civil Court as establishing a valid local custom under which the *zamindar* could recover from the *ryots* the whole or the part of the *patwari's* remuneration.

2.   Upon both these points the Board feel considerable doubt. The difficulty raised by the Commissioner and the Judge appears to be that in some villages the *ryots* are bound (or are alleged to be bound) by custom to pay to the *zamindar* the whole or a part of the *patwari's* salary.   The Court treat this claim as a demand for an illegal cess and refuse to decree the amount.   The *zamindar* is still liable to pay the *patwari*, but he is unable to make the *ryots* contribute.   The Commissioner proposes that a proceeding should be drawn up stating what the custom in each village is, and he considers that if a copy of this proceeding were attached to the plaint in a civil suit brought by the *zamindar* to recover from the *ryots* the contribution payable for the *patwari*, the Courts would recognize this proceeding as evidence of what Section 18 of the Regulation calls the mode of payment, and would decree the *zamindar's* claim.

3.   But the Board are inclined to think that the mode of payment mentioned in Section 18, refers only to the distinction between payment in money, in grain or in land, and has nothing to do with the question of the parties liable to pay the amount.   The whole tenor of the Regulation seem to imply that the remuneration of *patwari* is a question for the Collector and not for the Civil Court.   Section 34 expressly prohibits the Court from taking cognizance of the complaint of a *patwari* that he has not received his remuneration, and the same principle would equally seem to prohibit a *zamindar* from suing his *ryots* in the Civil Court for contribution towards the *patwari's* pay.

4.   The Board's records do not show that any orders under Sections 18, 19, and 20 of the Regulation have ever been issued in any district for determining the mode of payment of the *patwaris*, and the Board cannot think that the present time is opportune for giving effect to a provision which has remained dormant during so many years.

No. 1382.

FROM

T. T. ALLEN, Esq.,
Superintendent and Remembrancer of Legal Affairs.

To

THE SECRETARY TO THE BOARD OF REVENUE,
LOWER PROVINCES.

DATED CALCUTTA, *the 26th Jany.,* 1887.

SIR,

IN reply to your No. 1A of the 13th instant, I have the honor to say that in my opinion Sections 18, 19, and 20 of Regulation XII of 1817, having never been acted upon, must now, after the lapse of (70) seventy years, be deemed obsolete, and no action can be taken under them. The relations of *zamindars* and *ryots* have, during this interval utterly changed. Such a proceeding, as it is proposed now for Collector or Deputy Collector, to draw up, would, I expect, be treated as simple nullity by the Civil Court. If the *zamindar* on the ground of custom claims to recover from his *ryots* the *patwari's* pay, he must prove the custom in the Civil Court. The Deputy Collector's proceeding which embodies merely his opinion would not even be evidence in such a case.

Appendix II.

THE BEHAR LIGHT HORSE.

THE "first Volunteer Regiment on the Indian Army List, 'The Behar Light Horse' of the present day, was organized as the 'Behar Mounted Rifle Corps' in September 1862 with the consent of Government." Fifty-three of the planters of Tirhoot having already formed themselves into a corps, they forthwith, on the 16th July 1862, forwarded the following memorial to the Government of Bengal, desiring incorporation, through the Commissioner of Patna :—

"From certain residents in the districts of Tirhoot and Chuprah, to the Commissioner of Behar,—(dated the 16th July 1862).

SIR,—We, the undersigned residents in the districts of Tirhoot and Chuprah, beg to forward, through the Magistrate of Tirhoot, "for submission to the Government of India," our most loyal application, to be enrolled as members of a Volunteer Corps.

During the mutiny of 1857 the want of such a corps within these districts was greatly felt. Had there been such a Volunteer force (as now proposed) available, the local officers might have been spared the disgrace of deserting their station, and the services of the Bengal Yeomanry Cavalry and part of the Naval Police might have been dispensed with.

Although we have no reason to apprehend a recurrence of a similar emergency, yet still we are desirous of forming ourselves into a corps, whose services hereafter might be found available to the State.

We have before us extract No. 712 from the Proceedings of the Right Hon'ble the Governor-General in Council, in the Home Department, under date the 18th March 1861.

And in accordance with those instructions, under paragraph 1st, we have the honour to submit that the Corps already numbers at present fifty-three men, forming a troop for these Districts, and which is rapidly increasing.

We have considered the details of the duties that might be required from us, and from our knowledge of the two Districts, the nature of the country, &c., we are of opinion that a Mounted Rifle Corps, armed with a Sword and Terry's Breech-loading Rifle, would be the most serviceable.

The designation of the Corps should be Soubah Behar Mounted Rifle Corps.

We beg to submit the Rules and Regulations for the guidance of the Corps, for the approval of the Executive Government, as also the Rules under which it is proposed to conduct the Drill of the Corps.

In reference to the 3rd paragraph of these instructions, we beg that a suitable Cavalry Drill Instructor may be appointed. We believe the complement of officers to the number of men already enrolled to be—

1 Captain Commandant.
2 Lieutenants.
1 Cornet, doing duty also as Adjutant.

We beg also to submit herewith the names of the Commandant and 2nd-in-command for the approval of His Excellency the Governor-General, as per extract No. 71, and the names of the remaining officers for the sanction of the Government of Bengal.

Lastly, we have respectfully to request that, should His Excellency be pleased to sanction the formation of this corps, instructions may be issued in the Military Department for the issue of arms and ammunition under the Resolution No. 257 of the 7th of February 1861, page 425 of the *Government Gazette.*

P. S.—Chumparun.—Another troop is being formed in the District of Chumparun, and *which already* numbers seventeen members.

Commandant.—J. Forlong.

Lieutenants.—C. T. Metcalfe, Tirhoot Troop, and F. Holloway, Chumparun Troop.

Cornet doing duty also as *Adjutant.*—F. Collingridge.

RULES.

1. The Corps to be called the Soubah Behar Mounted Rifle Corps.

2. It consists of *effective* members only, and the number is unlimited.

3. It is armed with the Terry's Breech-loading Rifle.

4. The Commissioned and Non-commissioned strength of the Corps is for the present as follows :—

1 Commandant.	1 Sergeant-Major.
2 Lieutenants.	3 Sergeants.
1 Cornet being also Adjutant.	5 Corporals.

5. The uniforms and appointments are—Light-grey Coat with red piping, and Breeches with Jack Boots; Helmet with blue Pugree ; Waist Belt; Frog; Cartouche Box; Cap Pocket; Revolver Case fitting on to Sword Belt.

6. The admission and expulsion of Members is managed by a Committee of all the commissioned officers.

7. Applications for enrolment are to be made to the Adjutant, who will bring them before the Committee of Officers.

8. No Member to resign the Corps without giving due notice, not until he has made over his arms, &c., to the Commandant.

9. The Commandant shall have full powers in all matters connected with the discipline, drill, and internal economy of the Corps.

10. Non-attendance without permission, breaches of discipline, &c., can be punished by small fines not exceeding 8 annas, to be awarded by the Commanding Officer only.

11. No Member of the Corps to appear on parade unless in uniform.

12. Non-Commissioned Officers and Privates while on parade or duty are to salute their officers.

13. Damage done to arms, &c., when not on duty, is to be made good by the Member in whose possession they may be.

14. Drill will commence on the 15th of October of each year and last till the 15th of March. Within these dates members will be liable to be called out after the 15th of March to the 15th of October.

The Drill Instructor will proceed to certain convenient localities in the Mofussil, where members, residing within a circle of ten miles, will have to attend.

15. For all ordinary expenses, such as clearing and protecting the Drill and Practice grounds, keeping Arms, Target, Bugles, &c., in repair, each Member to pay a monthly subscription of Rs. 2.

16. A copy of these rules to be furnished to each Member of the Corps.

(Sd.) James Forlong."

To this loyal petition the Government replied in terms of sanction and compliment, under date 17th September 1862; the only item in which the Rules forwarded did not obtain the unqualified sanction of the Government was the Terry's Rifle, which, not being in store, the Corps were supplied with Sharpe's Breech-loading Carbine.

In 1870, the Officer Commanding the Corps brought to the notice of Government that there was a strong feeling amongst members in favour of carrying the Lance instead of the Arms which they had hitherto borne. To this the Secretary to the Government of India replied through the Bengal Government as follows :—

"From Colonel H. K. Burne, Secretary to the Government of India, Military Department, to the Secretary to the Government of Bengal, General Department,—(No. 513, dated Simla, the 26th July 1870.)

Sir,—With reference to your letter No. 655, dated 9th March 1870, bringing to notice the wish of the Members of the Behar Mounted Volunteer Rifle Corps to be allowed to carry Lances instead of their present Arms, I am directed to acquaint you, for the information o His Honor the Lieutenant-Governor, that the Right Hon'ble the Governor-General in Council does not consider the proposed change would at all conduce to the efficiency of the Corps. His Excellency in Council considers that the Behar Volunteers are much better armed, as at present, with Enfield carbines and swords than they would be with lances and swords, as, if ever they are called upon to act, it will be very possibly as infantry, and not as cavalry, and lances would, under these circumstances, be useless. Volunteers are much more useful for defence than for any other purpose, and therefore His Excellency the Governor-General in Council, with every desire to meet their wishes, thinks it best that the Behar Mounted Volunteers should retain their present weapons.

2. I am further directed to inform you that the rank of Major attaches to the post of Commandant, as notified in the letter from this Department, No. 777, dated 15th March 1865, and the necessary corrections will be made in the Army List, and the rank duly inserted in the commission to be sent to Major Collingridge."

We shall not enter into any inquiry as to the relative fairness from a military standpoint, of the wishes of the " Rifles " and the fiat of the Governor-General in Council ; it is enough to say that the corps had, at that time, acquired sufficient weight to induce Government to confer upon its Commanding Officer the rank of Major. Eight years had gone by since its formation, during which short time the strides made were truly wonderful. Evidently the Planters of Behar were growing in numbers, and, as they grew, so would their ardour to make their body of Volunteers the finest Corps in India intensify. Under the soldier-like command of Major Collingridge, the Behar Mounted· Rifles went on improving and increasing, and, when that fine old gentleman laid down his charge on his retirement Home, Major Hudson might well feel proud of the Corps which he had the honour to be elected to command. Nor was the new Commandant an officer at all likely to let the grass grow under his feet. He threw his whole soul into the work of organizing his Corps—now a regiment—as efficient in drill, arms, and practice, as it is possible for Volunteers to become. Ably assisted by Major Vousden, V.C., as his Adjutant, Major (now Lieutenant-Colonel) Hudson brought his regiment to a rare pitch of excellency, and its number rose from 120 to 320 sabres. The admirable work done by Major Vousden, who loved the Regiment as if they were his own Regulars, cannot be over-estimated.

Great changes have taken place since 1862. In the comparatively short space of 24 years, the arms, accoutrements, and uniform of this splendid body of horse, have undergone a complete revolution. The reader has already seen the regulation uniform, &c., which was worn by the Corps commanded by Captain Forlong in 1862. Through many different and indifferent variations, adaptations, and re-adaptations, the Regiment has now adopted the following uniform, &c., which, we presume, is permanent for some good time to come.

The uniform of the present is :—

Blue Blouse, Breeches, and Blucher Boots, white Pugree and Regulation Helmet with silver spike and chain, black and brown Belt, Martini-Henry Carbine and light Sword.

The Governments of India and Bengal have always displayed the keenest interest in the welfare and progress of this Regiment. The Lieutenant-Governor is Honorary Colonel *ex-officio*, and has often done them the compliment of coming to inspect them ; nay, it was only the great pressure of work and the urgent necessity for his presence in Upper Burma which prevented the Commander-in-Chief from paying the Behar Light Horse a similar well-merited compliment after the Camp of Exercise at Delhi. We do not know that any former Commander-in-Chief has ever expressly conveyed his desire to personally inspect this or any other Volunteer Regiment, and high compliment though it would undoubtedly have Leen, the Light Horse of Behar were well worthy of it, nor do they despair of mustering again ere long to acquit themselves like men and soldiers under the eye of one of the finest of England's Generals. When he comes, he will be thrice welcomed to Behar, and to the hearts of the loyal Behar Light Horse.

The North and South Ganges Troops inclusive, the Regiment now under the command of Lieutenant--Colonel Hudson, consists of troop, and musters 320 strong, and are, as we have before remarked, in a state of high efficiency.

But what boots it that so much has been done, whilst any remains still unaccomplished ? We have troops of the Light Horse having the following head-quarters :—Mozufferpore, Motihari, Chuprah, Gya, Arrah, and Bankipore, and not at one of these places have we a single fort or anything in the nature of one, which could afford adequate, or indeed the slightest, protection in times of trouble. This is surely not as it should be. Have we again lapsed into that false and deadly feeling of security and *laissez faire* which preceded the mutiny and its horrors? God forbid. A grain amongst the sands of the seashore—a drop in the vast ocean—a tiny bit of leaven amidst the lump of India's millions, so tiny that we may hardly hope and must not trust to leaven the whole lump—are we to disregard all due precaution and to be in jeopardy even to the eleventh hour, because the sky is clear, and all seems tranquil around us ? As the darkest night precedes the dawn, so also does the deepest calm forbode the storm. We are not pessimists,—all we would urge is that we are a small handful of British folk in a foreign land, amongst the teeming populations of which, with friendship there is mixed up hate, with fairness fanaticism, and with justice intolerance.

It behoves us in our Mofussil stations, at least in such of them as are non-military ones, to adopt all wise and necessary measures of precaution and defence. Defence, not attack, will always be the role of Volunteers, and, if this be so, then surely should there be places to defend. At the present moment there are absolutely none. Our helpless ones are at the mercy of any overwhelming force which might vanquish our Volunteer soldiers in the open field. This is not as it should be. At a very immaterial constructive cost, forts, such as we propose, could be established at all the abovenamed stations, and security with honour is cheap at *whatever* cost it may be purchased. We offer the suggestion, and hope to see its fulfilment at no far distant day.

We are not sufficiently masters of the technical knowledge of the craft to offer suggestions on our own account which would naturally occur to a soldier on seeing the material of the Regiment and the country which, were their services required, would be the probable scene of their campaigns.

We would be guilty of remissness did we pass without brief notice the sister Corps of Ghazipur Volunteers, commanded by Lieutenant-Colonel Rivett-Carnac. To this gentleman is due the sincere thanks of the whole British Indian community for his strenuous and indefatigable efforts to promote the Volunteer movement.

In October 1884 Mr. Rivett-Carnac submitted, at the request of the Lieutenant-Governor of the North-Western Provinces, proposals "for increasing the force of armed Europeans and Eurasians in the several Provinces of India, and for working thereto a reserve of native pensioned soldiers, to be employed as orderlies, &c., under the civil Governments." Mr. Rivett-Carnac, after adverting to the fact that there are 72,000 adult European and Eurasian males, estimated as capable of bearing arms in India and British Burma, goes on to show that, at the close of 1883, "there were then in India and Burma 12,421 Volunteers scattered over the country, divided into 49 corps." He then proceeds to show the great progress of the Volunteers in the North-Western Provinces, making reference to the "excellent example of the Behar Light Horse," and he represents the difficulties to be encountered in arming the Europeans at the last moment, after which he ably submits his proposals, of which we extract his recapitulation in full : —

" *Recapitulation* 24.—The recommendations herein contained may be recapitulated as follows : that—

(1). It is absolutely necessary to increase the armed European force in the country.

(2). The best way to do this would be to raise a Militia.

(3). Failing this a Register should be prepared by the civil authorities in every district, showing in detail all Europeans and Eurasians.

(4). From this a second Register should be prepared, showing the adult males capable of bearing arms.

(5). The local civil authorities should encourage all these to join the Volunteers or " Landwehr."

(6). Those who would not join the " Landwehr" should be invited to join a reserve or " *Landsturm.*"

(7). For the members of this Reserve, arms and ammunition should be supplied, and kept under proper precautions at the head-quarters of the districts.

(8). The members of the Reserve should be allowed, under certain restrictions, to keep and use their arms, and should be encouraged to join in the Volunteer Rifle matches, &c.

(9). The general control of the organization should be with the local civil officers, aided by the Volunteer Staff of Adjutants and Sergeants ; the Adjutant forming a link between the Commissioner and the General Commanding the Division.

(10). Full information regarding the number of men available, &c., in each district, should be furnished to the Military authorities, who would work with the civil administration in the matter.

(11). No efforts should be spared to render efficient the poorer classes of Eurasians, who require more drilling than the Europeans.

(12). For small bodies of European Riflemen in the districts, the regulations regarding drill should be relaxed, and the Boer organization followed ; Volunteer Corps at the large stations remaining as they are, with uniforms, drill, &c., after the manner of the Regular Army.

(13). The formation of parties of Light Horse or Uhlains should be encouraged in the district, thereby securing the best class of men as Volunteers.

(14.) District "meets" or social gatherings, with races, polo, and rifle-matches, should be encouraged and supported by the Government."

(15.) The heads of all departments of the Government should be induced to realize that the movement, being meritorious and necessary, should receive their active support.

(16.) And that certain privileges should be granted to members of existing Volunteer Corps."

The Behar Light Horse will not be likely soon to forget the reception they received on their visit to Calcutta at the invitation of their Honorary Colonel, Sir Rivers Thompson. On the 27th of December 1883, this gallant little Regiment arrived at the Howrah Railway Station, and, forming line, marched through the streets of Calcutta, to the inspiring notes of the Band of the 6th Warwickshire Regiment, on to their Barracks and Camp of Exercise at Ballygunge. Crowds accompanied them, a proof, if needed of the high estimation in which merry men of Behar are held by the inhabitants of the festive City of Palaces. They were, indeed, the heroes of the day as they rode along fully equipped and looking every man a soldier.

The Regiment lay encamped at Ballygunge for over a week, during which time they went through the exercises and drills, in fact the daily routine of the Regular Cavalry soldier.

The Behar Light Horse took part along with the Calcutta Mounted Rifles, the Calcutta Naval and City Volunteers, and the Regular Troops in the Proceedings on Proclamation day, and also in a sham-fight. Again, on the 4th of January, they, with the other Volunteers and the Troops in garrison, were paraded at the Brigade Parade Ground when they were inspected by His Excellency the Commander-in-Chief.

That fine cavalry soldier, General Wilkinson, Commanding in Calcutta, superintended the operations of the shamfight, and was pleased to compliment the Behar Regiment on their neat and soldier-like appearance, and the intelligent way in which they had performed their part in the day's proceedings.

It is noteworthy that all the arrangements connected with transport, commissariat mess, tents, &c., were planned by the Members of the Regiment, and were carried out by the Regimental Quarter-master and Officers without a single hitch. This is all the more creditable when we remember that the Regiment had never been massed together before. All were happy, and few were to be found amongst them who could say on his return that he had had a bad time.

R., In.   L

When we consider the sacrifices, not only of convenience and time, but also of "coin of the realm," which most of the members of this corps make, and many of its most efficient members, young Indigo Planters, are not overburdened by the latter. It is a marvel how so many men manage to attend every Troop-drill Muster. When we speak of them, "Attending Parade," outsiders do not know that in this case it means that a man has travelled 20, 30, and even 100 miles across country before he turns up mounted.

When the South Ganges Troop was being raised, the writer of this book was one of the chief promoters, and rode in from Rohtasghur to Arrah, a distance very little short of 100 miles. This, too, I did on one of the hottest days of a hot May. We celebrated the Queen's Birthday right royally, and the 24th May 1882, I think, saw the formation of the latest addition to the Behar Light Horse, the South Ganges Troop.

I went in with proxies in my pocket and carried the men I wanted as Captain and Lieutenants, men whom I knew would steer the infant-troop to the present high position it now holds in the Regiment.

When every honest British heart was wrung by the tidings of the sad stress in which the hero Gordon was placed, Colonel Hudson called on his six troops' leaders for Volunteers for the Soudan, and promptly received answer that a hundred men and horses were ready. But alas ! ere the tender of their services reached the Viceroy, Gordon had fallen. It must have added, if possible, to the bitter regret for his cruel fate felt by every one of these good men, and true to find they were also " too late " to rescue or share his fate.

The Viceroy, in conveying the thanks of Government to the Regiment, expressed strongly his appreciation of the value of its services, and of its gallant offer of them. There is a muster *de rigeur* once a month at each Troop Head-quarters for two or three days' drill and practice on the range, and at certain seasons, often in the hottest weather, as then it is that the Planter can spare the time ; the Troops muster twice a month. It is only a strong soldierly instinct and thorough *esprit de corps* that give such a result ; when also do men perform such duties for years for honor alone, and with the simple desire that were they called on they may be ready to do credit to their manhood.

𝔄𝔭𝔭𝔢𝔫𝔡𝔦𝔵 𝕴𝕴𝕴.

TABLE I.

An account of Indigo, manufactured in Bengal and its dependant provinces, imported into Calcutta annually, from 1795-96 *to* 1831-32.

Years.	Factory Mds.	Years.	Factory Mds,
1795-96	62,500	1814-15	102,662
1796-97	32,300	1815-16	114,481
1797-98	54,600	1816-17	83,000
1798-99	23,800	1817-18	72,000
1799-1800	35,540	1818-19	75,000
1800-1	39,900	1819-20	106,843
1801-2	38,500	1820-21	76,?54
1802-3	29,800	1821-22	92,848
1803-4	54,048	1822-23	112,606
1804-5	64,803	1823-24	80,315
1805-6	85,380	1824-25	110,227
1806-7	51,244	1825-26	156,548
1807-8	103-950	1826-27	79,678
1808-9	94,539	1827-28	151,699
1809-10	43,012	1828-29	98,009
1810-11	73,407	1829-30	132,946
1811-12	69,654	1830-31	129,117
1812-13	73,883	1831-32	121,000
1813-14	74,585	....	......

Indigo Calcutta Price Current.

	1823	1824	1825	1826	1827	1828	1829
In February & December.							
Fine blue	....	325@335 295,,300	300@310 260,,270	...... 295@300	0@ 0 270,,290	290@310 290,,310	285@310 ......
Ordinary ditto	....	315,,320 280,,290	285,,295 240,,250	...... 280@290	280,,300 250,,265	255,,280 270,,285	270@280 ......
Fine purple and violet	290@295	305,,315 280,,290	285,,295 250,,260	...... 285@290	290,,300 250,,260	255,,265 270,,290	270@280 ......
Ordinary ditto	280@285	260,,300 260,,270	260,,775 230,,240	...... 260@270	265,,280 235,,245	240,,250 250,,260	255@265 ......
Dull blue	260@270	240,,260	240,,260 200,,210	......	240,,250	220,,235	......
Inferior purple & violet	240@250	250,,260	220,,235	......	230@240	200@220	......
Strong copper	275,,285	270,,290	210,,220	......	230,,240	200,,200	......
In February & July.							
Ordinary copper	230@240	260@280 200,,220	190@200 190,,200	....	180@220 160,,240	180@220	....
Oude fine	250@260	210,,240 130,,160	180,,199 180,,190	....	280,,285 160,,240	200@250	....
Ditto ordinary	200@229	180,,200 65,,120	60,,160 60,,160	....	100,,180 60,,120	100@180	....

Indigo Calcutta Price Current.

YEARS AND MONTHS.	Blue.	Blue and purple.	Purple.	Purple and violet.	Violet.	Violet and copper.	Fine copper.
	P. md.	P. md.	P. md.	P. md.	P. md.	P. md.	P. md.
	Sa. Rs.	Sa. Rs.	Sa. Rs.	Sa. Rs.	Sa. Rs.	Sa. Rs.	Sa. Rs.
1815 January	200	185	175	150	135	120	110
1815 April	185	170	160	145	125	120	110
1815 July	0	155	145	130	125	120	110
1816 January	0	160	155	150	140	135	130
1816 April	0	0	140	135	130	125	120
1816 July	0	0	140	135	130	125	120
1817 January	160	150	145	140	135	130	120
1817 April	0	0	0	140	135	130	120
1817 July	0	0	0	0	0	130	120
1818 January	175	165	150	140	135	130	120
1818 April	0	0	0	0	135	130	120
1818 July	0	0	0	0	0	0	0
1819 January	180	175	170	160	150	140	135
1819 April	185	170	160	155	145	138	130
1819 July	0	0	0	0	0	0	5
1820 January	160	150	145	140	130	120	0
1820 April	0	0	0	140	130	120	105
1820 July	0	0	0	0	0	0	0
1821 January	0	0	175	170	160	150	145
1821 April	0	0	175	170	160	150	145
1821 July	0	0	0	0	·0	0	0
1822 January	270	260	245	215	195	190	210
1822 April	0	0	240	215	195	190	210
1822 July	0	0	0	247	220	195	215
1822 October	320	0	0	290	0	0	0

1887.

SELECTED LIST

OF

Illustrated and General Publications

BY

THACKER, SPINK & CO., CALCUTTA.

W. THACKER & CO., 87, NEWGATE ST., LONDON.

The Tribes on My Frontier: an Indian Naturalist's Foreign Policy. By EHA. With 50 Illustrations by F. C. MACRAE. In Imperial 16mo. Uniform with "Lays of Ind," "Riding," "Hindu Mythology," &c. Third Edition, Rs. 5-8 (8s. 6d.)

This remarkably clever work most graphically and humorously describes the surroundings of a country bungalow. The twenty chapters embrace a year's experiences, and provide endless sources of amusement and suggestion. The numerous able illustrations add very greatly to the interest of the volume, which will find a place on every table.

"It is a very clever record of a year's observations round the bungalow in 'Dustypore.' . . . . It is by no means a mere travesty. . . . . The writer is always amusing, and never dull."—*Field.*

"The book is cleverly illustrated by Mr. F. C. Macrae. We have only to thank our Anglo-Indian naturalist for the delightful book which he has sent home to his countrymen in Britain. May he live to give us another such."—*Chambers' Journal.*

"A most charming series of sprightly and entertaining essays on what may be termed the fauna of the Indian bungalow. . . . . . We have no doubt that this amusing book will find its way into every Anglo-Indian's library."—*Allen's Indian Mail.*

"This is a delightful book, irresistibly funny in description and illustration, but full of genuine science too. . . . . . There is not a dull or uninstructive page in the whole book."—*Knowledge.*

"It is a pleasantly-written book about the insects and other torments of India which make Anglo-Indian life unpleasant, and which can be read with pleasure even by those beyond the reach of the tormenting things 'Eha' describes."—*Graphic.*

"The volume is full of accurate and unfamiliar observation, and the illustrations prove to be by no means without their value."—*Saturday Review.*

A Natural History of the Mammalia of India,

Burmah and Ceylon. By R. A. STERNDALE, F.R.G.S., F.Z.S., &c., Author of "Seonee," "The Denizens of the Jungle," "The Afghan Knife," &c. With 170 Illustrations by the Author and Others. In Imperial 16mo. Uniform with "Riding," "Hindu Mythology," and "Indian Ferns." Rs. 10. (12s. 6d.)

"It is the very model of what a popular natural history should be."
—*Knowledge.*

"An amusing work with good illustrations."—*Nature.*

"Full of accurate observation, brightly told."—*Saturday Review.*

"The results of a close and sympathetic observation."—*Athenœum.*

"It has the brevity which is the soul of wit, and a delicacy of allusion which charms the literary critic."—*Academy.*

"The notices of each animal are, as a rule, short, though on some of the larger mammals—the lion, tiger, pard, boar, &c.—ample and interesting details are given, including occasional anecdotes of adventure. The book will, no doubt, be specially useful to the sportsman, and, indeed, has been extended so as to include all territories likely to be reached by the sportsman from India. . . . Those who desire to obtain some general information, popularly conveyed, on the subject with which the book deals, will, we believe, find it useful."—*The Times.*

"Has contrived to hit a happy mean between the stiff scientific treatise and the bosh of what may be called anecdotal zoology."—*The Daily News.*

Handbook to the Ferns of India, Ceylon, and

the Malay Peninsula. By Colonel R. H. BEDDOME, Author of the "Ferns of British India," "The Ferns of Southern India." Three hundred Illustrations by the Author. Uniform with "Lays of Ind," "Hindu Mythology," "Riding," "Natural History of the Mammalia of India," &c. Imperial 16mo. Rs. 10. (12s. 6d.)

"The great amount of care observed in its compilation makes it a most valuable work of reference, especially to non-scientific readers ; for, in preparing it, as many of the technicalities as could be safely dispensed with are left aside. . . . . . A magnificent volume of nearly 500 pages, illustrated with 300 admirable woodcuts."—*Garden.*

"It is the first special book of portable size and moderate price which has been devoted to Indian Ferns, and is in every way deserving of the extensive circulation it is sure to obtain."—*Nature.*

"Will prove vastly interesting, not only to the Indian people, but to the botanists of this country."—*Indian Daily News.*

"This is a good book, being of a useful and trustworthy character. The species are familiarly described, and most of them illustrated by small figures."—*Gardeners' Chronicle.*

"Those interested in botany will do well to procure a new work on the 'Ferns of British India.' The work will prove a first-class text-book."—*Free Press.*

Lays of Ind. By ALIPH CHEEM. Comic, Satirical, and Descriptive Poems illustrative of Anglo-Indian Life. Seventh Edition. Enlarged. With 70 Illustrations. Cloth elegant, gilt edges. Rs. 7 (10s. 6d.)

"Aliph Cheem presents us in this volume with some highly amusing ballads and songs, which have already in a former edition warmed the hearts and cheered the lonely hours of many an Anglo-Indian, the pictures being chiefly those of Indian life. There is no mistaking the humour, and at times, indeed, the fun is both 'fast and furious.' One can readily imagine the merriment created round the camp fire by the recitation of 'The Two Thumpers,' which is irresistibly droll. . . . The edition before us is enlarged, and contains illustrations by the author, in addition to which it is beautifully printed and handsomely got up, all which recommendations are sure to make the name of Aliph Cheem more popular in India than ever."—*Liverpool Mercury.*

"The 'Lays' are not only Anglo-Indian in origin, but out-and-out Anglo-Indian in subject and colour. To one who knows something of life at an Indian 'station' they will be especially amusing. Their exuberant fun at the same time may well attract the attention of the ill-defined individual known as the 'general reader.'"—*Scotsman.*

"This is a remarkably bright little book. 'Aliph Cheem, supposed to be the *nom de plume* of an officer in the 18th Hussars, is, after his fashion, an Indian Bon Gaultier. In a few of the poems the jokes, turning on local names and customs, are somewhat esoteric; but taken throughout, the verses are characterised by high animal spirits, great cleverness, and most excellent fooling."—*World.*

"To many Anglo-Indians the lively verses of 'Aliph Cheem' must be very well known; while to those who have not yet become acquainted with them we can only say, read them on the first opportunity. To those not familiar with Indian life they may be specially commended for the picture which they give of many of its lighter incidents and conditions, and of several of its ordinary personages."—*Bath Chronicle.*

"Satire of the most amusing and inoffensive kind, humour the most genuine, and pathos the most touching pervade these 'Lays of Ind.' . . . From Indian friends we have heard of the popularity these 'Lays' have obtained in the land where they were written, and we predict for them a popularity equally great at home."—*Monthly Homœopathic Review.*

"Former editions of this entertaining book having been received with great favour by the public and by the press, a new edition has been issued in elegant type and binding. The Author, although assuming a *nom de plume*, is recognised as a distinguished cavalry officer, possessed of a vivid imagination and a sense of humour amounting sometimes to rollicking and contagious fun."—*Capital and Labour.*

Riding : On the Flat and Across Country. A Guide to Practical Horsemanship. By Capt. M. H. HAYES. Illustrated by Sturgess. Second Edition. Revised and Enlarged. Imperial 16mo. Rs. 7 (10s. 6d.)

"The book is one that no man who has ever sat in a saddle can fail to read with interest."—*Illustrated Sporting and Dramatic News.*

"An excellent book on riding."—*Truth.*

"Mr. Hayes has supplemented his own experience on race-riding by resorting to Tom Cannon, Fordham, and other well-known jockeys for illustration. 'The Guide' is, on the whole, thoroughly reliable ; and both the illustrations and the printing do credit to the publishers."—*Field.*

"It has, however, been reserved for Captain Hayes to write what in our opinion will be generally accepted as the most comprehensive, enlightened and 'all round' work on riding, bringing to bear as he does not only his own great experience, but the advice and practice of many of the best recognised horsemen of the period."—*Sporting Life.*

"Captain Hayes is not only a master of his subject, but he knows how to aid others in gaining such a mastery as may be obtained by the study of a book."—*The Standard.*

COMPANION VOLUME TO THE ABOVE.

Riding for Ladies, with Hints on the Stable. A Lady's Horse Book. By Mrs. POWER O'DONOGHUE. Author of "A Beggar on Horseback," "Ladies on Horseback," "Unfairly Won," &c. With 91 Illustrations, by A. CHANTREY CORBOULD, and portrait of the Author. Elegantly printed and bound. Imperial 16mo. Rs. 10. (12s. 6d.)

I.—Ought Children to Ride?
II.—"For Mothers and Children."
III.—First Hints to a Learner.
IV.—Selecting a Mount.
V., VI.—The Lady's Dress.
VII.—Bitting. VIII.—Saddling.
IX.—Sit, Walk, Canter, and Trot.
X.—Reins, Voice, and Whip.
XI.—Riding on the Road.
XII.—Paces, Vices, and Faults.
XIII.—A Lesson in Leaping.
XIV.—Managing Refusers.
XV.—Falling.
XVI.—Hunting Outfit Considered.
XVII.—Economy in Riding Dress.
XVIII.—Hacks and Hunters.
XIX.—In the Hunting Field.
XX.—Shoeing. XXI.—Feeding.
XXII.—Stabling. XXIII.—Doctoring
XXIV.—Breeding. XXV.—"Tips."

"Mrs. Power O'Donoghue (more power to her—not that she wants it) shows no sign of 'falling off.' Indeed, she shows her readers how to become riders, and to stick on gracefully. She sketches her pupils 'in their habits as they ride,' and gives them a bit of her mind about bits, and tells them about spurs on the spur of a moment."—*Punch.*

"Mrs. O'Donoghue is great on the subject of a lady's riding-dress, and lays down some useful information which should not be forgotten. . . . . From first to last she never errs on the side of anything approaching to bad taste, which is more than can be said for some equestriennes."—*Field.*

"It is characteristic of her book, as of all books of any value, that it has a distinctive character. Sound common sense, and a thoroughly practical way of communicating instruction, are its leading traits."—*Daily News.*

Splendidly Illustrated book of Sport. In Demy 4to ; Rs. 25 ; elegantly
bound. (£2 2s.)

Large Game Shooting in Thibet, the Hima-
layas, and Northern India. By Colonel ALEXANDER A. KIN-
LOCH. Containing descriptions of the country and of the
various animals to be found ; together with extracts from a
journal of several years' standing. With thirty illustrations
and map of the district.

"An attractive volume, full of sporting adventures in the valleys and
forest hills extending along the foot of the Himalayas. Its pages are also
interesting for the graphic description they give of the beasts of the field,
the cunning instinct which they show in guarding their safety, the places
which they choose for their lair, and the way in which they show their
anger when at bay. Colonel Kinloch writes on all these subjects in a
genuine and straightforward style, aiming at giving a complete description
of the habits and movements of the game."—*British Mail.*

"If Carlyle had ever condescended to notice sport and sportsmen he might
probably have invented some curious and expressive phrase for the author
of this book. It is the work of a genuine shikari . . . The heads have
been admirably reproduced by the photograph. The spiral or curved horns,
the silky hair, the fierce glance, the massive jaws, the thick neck of deer,
antelope, yak or bison, are realistic and superior to anything that we can
remember in any bookshelf full of Indian sport."—*Saturday Review.*

"The splendidly illustrated record of sport. The photo-gravures, especially the heads of the various antelopes, are lifelike ; and the letterpress
is very pleasant reading."—*Graphic.*

Denizens of the Jungles ; a series of Sketches of Wild
Animals, illustrating their form and natural attitude. With
letterpress description of each plate. By R. A. STERNDALE,
F.R.G.S., F.Z.S. Author of "Natural History of the Mammalia of India," " Seonee," &c. Oblong folio. Rs. 10. (16s.)

I.—Denizens of the Jungles. Aborigines — Deer — Monkeys.	VII.—"A Race for Life." Blue Bull and Wild Dogs.
II.—"On the Watch." Tiger.	VIII.—"Meaning Mischief." The Gaur—Indian Bison.
III.—"Not so Fast Asleep as he Looks." Panther—Monkeys.	IX.—"More than His Match.' Buffalo and Rhinoceros.
IV.—"Waiting for Father." Black Bears of the Plains.	X.—"A Critical Moment. Spotted Deer and Leopard.
V.—"Rival Monarchs." Tiger and Elephant.	XI.—"Hard Hit." The Sambur.
VI.—"Hors de Combat." Indian Wild Boar and Tiger.	XII.—"Mountain Monarchs." Marco Polo's Sheep.

Useful Hints to Young Shikaris on the Gun and
Rifle. By "THE LITTLE OLD BEAR." Reprinted from
the *Asian.* Crown 8vo. Rs. 2-8.

Third Edition, revised, enlarged, and newly Illustrated.
Crown 8vo. Rs. 7. (10s. 6d.)

Veterinary Notes for Horse-Owners.—An everyday Horse Book. By Captain M. HORACE HAYES, M.R.C.V.S.

OPINIONS OF THE PRESS

"The work is written in a clear and practical way."—*Saturday Review.*

"Of the many popular veterinary books which have come under our notice, this is certainly one of the most scientific and reliable. . . . Some notice is accorded to nearly all the diseases which are common to horses in this country, and the writer takes advantage of his Indian experience to touch upon several maladies of horses in that country, where veterinary surgeons are few and far between. The description of symptoms and the directions for the application of remedies are given in perfectly plain terms, which the tyro will find no difficulty in comprehending : and, for the purpose of further smoothing his path, a chapter is given on veterinary medicines, their actions, uses, and doses."—*The Field.*

"Simplicity is one of the most commendable features in the book. What Captain Hayes has to say he says in plain terms, and the book is a very useful one for everybody who is concerned with horses."—*Illustrated Sporting and Dramatic News.*

"We heartily welcome the second edition of this exceedingly useful book. The first edition was brought out about two years since, but the work now under notice is fully double the size of its predecessor, and, as a matter of course, contains more information. Captain Hayes, the author, is not only a practical man in all things connected with the horse, but has also studied his subject from a scientific point of view."—*The Sporting Life.*

"Captain Hayes, in the new edition of 'Veterinary Notes,' has added considerably to its value by including matter which was omitted in the former editions, and rendered the book, if larger, at any rate more useful to those non-professional people who may be inclined or compelled to treat their own horses when sick or injured. So far as we are able to judge, the book leaves nothing to be desired on the score of lucidity and comprehensiveness."—*Veterinary Journal.*

"Captain Hayes has succeeded in disposing of two editions of his manual since it was issued in 1877—a sufficient proof of its usefulness to horse-owners. The present edition is nearly double the size of the first one, and the additional articles are well and clearly written, and much increase the value of the work. We do not think that horse-owners in general are likely to find a more reliable and useful book for guidance in an emergency."—*The Field.*   New Edition, Revised.

Training and Horse Management in India. By Captain M. HORACE HAYES, author of "Veterinary Notes for Horse Owners," "Riding," &c. Third Edition. Crown 8vo. Rs. 5. (8s. 6d.)

"No better guide could be placed in the hands of either amateur horseman or veterinary surgeon."—*The Veterinary Journal.*

"A useful guide in regard to horses anywhere. . . . . . Concise, practical, and portable."—*Saturday Review.*

Indian Notes about Dogs: their Diseases and Treatment. By Major C——. Third Edition, Revised. Fcap. 8vo., cloth. Re. 1-8.

Indian Racing Reminiscences. Being Entertaining Narratives and Anecdotes of Men, Horses, and Sport. By Captain M. HORACE HAYES, Author of "Veterinary Notes," "Training and Horse Management," &c. Illustrated with 22 Portraits and 20 Engravings. Imperial 16mo. Rs. 5-12. (8s. 6d.)

"Captain Hayes has done wisely in publishing these lively sketches of life in India. The book is full of racy anecdote."—*Bell's Life.*

"All sportsmen who can appreciate a book on racing, written in a chatty style, and full of anecdote, will like Captain Hayes's latest work."—*Field.*

"It is a safe prediction that this work is certain to have a wide circle of readers."—*Broad Arrow.*

"The book is valuable from the fact that many hints on the treatment of horses are included, and the accuracy and extent of Captain Hayes's veterinary skill and knowledge are well known to experts."—*Illustrated Sporting and Dramatic News.*

"Many a racing anecdote and many a curious character our readers will find in the book, which is very well got up, and embellished with many portraits."—*Baily's Magazine.*

Hindu Mythology: Vedic and Puranic. By Rev. W. J. WILKINS, of the London Missionary Society, Calcutta. Illustrated by very numerous Engravings from Drawings by Native Artists. Uniform with "Lays of Ind," "Riding," &c. Rs. 7. (10s. 6d.)

"His aim has been to give a faithful account of the Hindu deities such as an intelligent native would himself give, and he has endeavoured, in order to achieve his purpose, to keep his mind free from prejudice or theological bias. To help to completeness he has included a number of drawings of the principal deities, executed by native artists. The author has attempted a work of no little ambition and has succeeded in his attempt, the volume being one of great interest and usefulness; and not the less so because he has strictly refrained from diluting his facts with comments of his own. It has numerous illustrations."—*Home News.*

"Mr. Wilkins has done his work well, with an honest desire to state facts apart from all theological prepossession, and his volume is likely to be a useful book of reference."—*Guardian.*

"In Mr. Wilkins's book we have an illustrated manual, the study of which will lay a solid foundation for more advanced knowledge, while it will furnish those who may have the desire without having the time or opportunity to go further into the subject, with a really extensive stock of accurate information."—*Indian Daily News.*

Modern Hinduism: Being an Account of the Religion and Life of the Hindus in Northern India. By W. J. WILKINS, of the London Missionary Society, Author of "Hindu Mythology, Vedic and Puranic." Demy 8vo. Price Rs. 8.

Indian Horse Notes: an Epitome of useful Information arranged for ready reference on Emergencies, and specially adapted for Officers and Mofussil Residents. All Technical Terms explained and Simplest Remedies selected. By Major C——, Author of "Indian Notes about Dogs;" Second Edition, Revised and considerably Enlarged. Fcap. 8vo. cloth. Rs. 2.

Eighth Edition.   Crown 8vo.   Rs. 7.   (10s. 6d.)

The Management and Medical Treatment of
Children in India.   By EDWARD A. BIRCH, M.D., Surgeon Major Bengal Establishment.   Second Edition, Revised. Being the Eighth Edition of "Goodeve's Hints for the Management of Children in India."

Dr. Goodeve.—"I have no hesitation in saying that the present edition is for many reasons superior to its predecessors. It is written very carefully, and with much knowledge and experience on the author's part, whilst it possesses the great advantage of bringing up the subject to the present level of Medical Science."

The Medical Times and Gazette, in an article upon this work and Moore's "Family Medicine for India," says :—The two works before us are in themselves probably about the best examples of medical works written for non-professional readers. The style of each is simple, and as free as possible from technical expressions. The modes of treatment recommended are generally those most likely to yield good results in the hands of laymen ; and throughout each volume the important fact is kept constantly before the mind of the reader, that the volume he is using is but a poor substitute for personal professional advice, for which it must be discarded whenever there is the opportunity.

A Tea Planter's Life in Assam.   By GEORGE M.
BARKER.   With Seventy-five Illustrations by the Author. Crown 8vo.   Rs. 5.   (7s. 6d.)

"Mr. Barker has supplied us with a very good and readable description, accompanied by numerous illustrations drawn by himself. What may be called the business parts of the book are of most value."—*Contemporary Review.*

"Cheery, well-written little book."—*Graphic.*

"A very interesting and amusing book, artistically illustrated from sketches drawn by the Author."—*Mark Lane Express.*

A Complete List of Indian Tea Gardens, Indigo
Concerns, Silk Filatures, Sugar Factories, Cinchona Concerns, and Coffee Estates. With their Capital, Directors, Proprietors, Agents, Managers, Assistants, &c., and their Factory Marks by which the chests may be identified in the market.   5s.

"The strong point of the book is the reproduction of the factory marks, which are presented side by side with the letterpress. To buyers of tea and other Indian products on this side, the work needs no recommendation."—*British Trade Journal.*

The Tea Estates of Ceylon, their Acreage and
Proprietors.   1s. 6d., or with the "Indian Tea Gardens," 6s.

Merces' Indian and English Exchange Tables

from 1s. 4d. to 1s. 8d. per Rupee. New Edition. In this Edition the rate rises by 32nds of a penny, to meet the requirements of Financiers. The progression of the numbers is by units ; thus, in most instances, saving a line of calculation. Accuracy, facility of reference, and perfectly clear printing, render it the most perfect work in existence. Demy 8vo. Rs. 10. (15s.)

Supplement containing 1/5 to 1/5$\frac{31}{32}$. Rs. 3-8. (5s.)
      ditto   1/4 to 1/4$\frac{31}{32}$. Rs. 3-8. (5s.)

"In this new edition of Mr. Merces' useful work, the calculations have been extended to thirty-seconds of a penny, and all sums from £1 to £100, and from 1 to 100 rupees, are made to advance by units."—*Economist.*

"We heartily recommend these tables, both for their reliability and for the great saving in time that will be gained by their employment."—*Financier.*

Our Administration of India : being a complete

Account of the Revenue and Collectorate Administration in all Departments, with special reference to the Work and Duties of a District Officer in Bengal. By H. A. D. PHILLIPS. Rs. 4-4. (6s.)

"In eleven chapters Mr. Phillips gives a complete epitome of the civil, in distinction from the criminal, duties of an Indian Collector. The information is all derived from personal experience. A polemical interest runs through the book, but this does not detract from the value of the very complete collections of facts and statistics given."—*London Quarterly Review.*

"It contains much information in a convenient form for English readers who wish to study the working of our system in the country districts of India."—*Westminster Review.*

"A very handy and useful book of information upon a very momentous subject, about which Englishmen know very little."—*Pall Mall Gazette.*

The Reconnoitrer's Guide and Field Book,

adapted for India. By Lieut.-Col. M. J. KING-HARMAN, B.S.C. Second Edition, Revised and Enlarged. In roan. Rs. 3.

It contains all that is required for the guidance of the Military Reconnoitrer in India : it can be used as an ordinary Pocket Note Book, or as a Field Message Book; the pages are ruled as a Field Book, and in sections, for written description or sketch.

"To officers serving in India this guide will be invaluable."—*Broad Arrow.*

Tales from Indian History: being the Annals of India retold in Narratives. By J. TALBOYS WHEELER. Crown 8vo., cloth gilt. Rs. 3-4. (5s.)

"No young reader who revolts at the ordinary history presented to him in his school books will hesitate to take up this. No one can read a volume such as this without being deeply interested."—*Scotsman.*

"While the work has been written for them (natives), it has also been written for the people of England, who will find in the volume, perhaps for the first time, the history of our great dependency made extremely attractive reading. Mr. Wheeler's narrative is written in a most careful style; indeed, he is master of the English language. He does not confine himself to the mere dry details of history, but tells the adventures of Indian heroes and heroines in legends of love and war; describes the village communities of India, their organization and self-government; delineates the results of caste, infant marriage, and other Hindu institutions and usages as seen in the family and social life of the people in villages and towns, as well as in courts and palaces. . . . The work also contains valuable observations on the foreign relations of the Indian Empire with Persia, Russia, Turkey, and China. Altogether this is a work of rare merit."—*Broad Arrow.*

"In going through an interesting book, the reader will be furnished with a good general notion of Indian history, and learn besides something about Indian modes of life."—*Queen.*

"Will absorb the attention of all who delight in thrilling records of adventure and daring. It is no mere compilation, but an earnest and brightly-written book."—*Daily Chronicle.*

"This little volume contains a history of India in the form of tales and narratives, intended by the author for the people of India as well as for those of the British Isles."—*Army and Navy Gazette.*

The Student's Manual of Tactics. By Capt. M. HORACE HAYES. Specially written for the use of candidates preparing for the Militia, Military Competitive Examinations, and for promotion. Crown 8vo. Rs. 4-4. (6s.)

Definitions.
I. Composition of an Army.
II. Infantry.
III. Artillery.
IV. Cavalry.
V. Formations: Time and Space.
VI. Outposts.
VII. Screening and Reconnoitring.
VIII. Advanced Guards.
IX. Rear Guards.
X. Marches.
XI. The Attack.
XII. The Defence.
XIII. Villages.
XIV. Woods.
XV. Machine Guns.

"There is no better Manual on Tactics than the one which Captain Hayes has written."—*Naval and Military Gazette.*

"'The Student's Manual of Tactics' is an excellent book. Principles are reasoned out, and details explained in such a way that the student cannot fail to get a good grasp of the subject. Having served in both the artillery and infantry, and being a practical writer, as well as 'a coach,' the author of this manual had exceptional qualifications for the task he has accomplished."—*Broad Arrow.*

UNDER PATRONAGE OF THE SECRETARY OF STATE.
In Royal 8vo. Rs. 22. (31s. 6d.)

Statistics of Hydraulic Works, and Hydrology

of England, Canada, Egypt, and India. Collected and reduced by LOWIS D'A. JACKSON, C.E., Author of "Canal and Culvert Tables," "Hydraulic Manual," "Aid to Engineering Solution," &c.

". . . The intention of the author being apparently to bring together in a compact and easily accessible form a mass of information, which is for the most part at present buried in official archives, or not readily attainable even to professional men. Though apparently compiled primarily for the benefit of the India Public Works' Department, the book contains much information which is not generally known in England even amongst engineers, especially as regards the gigantic scale on which hydraulic works are carried out in foreign countries. Mr. Jackson's book concludes with a short account of the works carried out in Ceylon."—*The Builder*.

"In this work a successful attempt has been made to collect and arrange in a systematic order facts and data. . . . The order in which the subjects are arranged being river basins : their area in square miles, geology and rainfall, canals and navigation, storage works, irrigation with sewage, analysis of water and of factory effluents, &c. The statistics relating to canals have been compiled with much care. Irrigation with sewage is a full and complete chapter. The analysis of water will be welcomed as a very useful summary. The hydrology of Canada, Egypt, and India, especially the latter, is very carefully tabulated. The latter portion of Mr. Jackson's laborious work will be of considerable value to engineers of the Indian Public Works' Department."—*Building News*.

A Manual of Surveying for India, detailing the

mode of operations on the Trigonometrical, Topographical and Revenue Surveys of India. Compiled by Sir H. L. THUILLIER, K.C.S.I., and Lieut.-Col. R. SMYTH. Prepared for the use of the Survey Department, and published under the authority of the Government of India. Royal 8vo. Rs. 16. (30s.)

The Hindoos as they are : a description of the Manners,

Customs, and Inner Life of Hindoo Society. Bengal. By SHIB CHUNDER BOSE. Second Edition. Revised. Crown 8vo. Rs. 5.

"Lifts the veil from the inner domestic life of his countrymen."—*Westminster Review*.

A Memoir of the late Justice Onoocool Chunder

Mookerjee. By M. MOOKERJEE. Third Edition. 12mo. Re. 1. (2s. 6d.)

The Biography of a Native Judge, by a native, forming a most interesting and amusing illustration of Indian English.

"The reader is earnestly advised to procure the life of this gentleman, written by his nephew, and read it."—*The Tribes on my Frontier*.

Hints on the Study of English. By F. J. Rowe,
M.A., and W. T. Webb, M.A., Professors of English Literature, Presidency College, Calcutta.   New Edition, Revised.
Crown 8vo., cloth.   Rs. 2-8.   1887.
This Edition has been carefully revised throughout, and contains
a large amount of new matter, specially adapted to the requirements
of Native Students and Candidates for University Examination.

"Messrs. Rowe and Webb have thoroughly grasped not only the relations between the English tongue and other tongues, but the fact that
there is an English tongue. . . . We are thoroughly glad to see native
Indian students of English taught the history and nature of our language
in a way in which, only a few years back, no one would have been taught
at home."—*Saturday Review.*

"In the work before us, Messrs. Rowe and Webb have produced what,
for the special purpose for which it is intended, viz.—the instruction of
native and Eurasian students—is by far the best manual of the English
language we have yet seen."—*Englishman.*

"So far as it goes, this is one of the most satisfactory books of the kind
that we have seen. No point touched upon is slurred over ; a great deal
of matter is condensed into a small compass, and at the same time expressed
in a simple, easy style. . . . Taking it as a whole, this is a scholarly little
work ; and, as such, its usefulness will not be limited to one small class of
students."—*Times of India.*

"I wish to say that the book shows wonderful toil and care, and is above
the average even for purely English readers : for the particular purpose, it
is, I should suppose, admirable."—*Extract from a letter from the Rev. W. W.
Skeat, M.A., Professor of Anglo-Saxon in the University of Cambridge.*

A Companion Reader to "Hints on the Study of English."
(Eighteenth Thousand.)   Demy 8vo.   Price Rs. 1-4.
"The passages selected are, in most cases, admirably adapted for the
purpose in view, and the notes generally give the student neither less than
he ought to expect, nor more than he ought to get."—*Englishman.*

"We have no hesitation in saying that Messrs. Rowe and Webb have
rendered excellent service to the cause of education in their selections and
their method of treating them for the purpose intended."—*Indian Daily News.*

"The authors of the 'Hints' have rendered an additional service to the
cause of English education, by supplying a 'Companion Reader,' of whose
merits it would not be easy to speak too highly. . . . It is not merely a
Reader, but a most suggestive and judicious guide to teachers and students."
—*Friend of India.*   Crown 8vo, Rs. 4. (7s. 6d.)

Indian Lyrics. By W. Trego Webb, M.A., Bengal
Education Service.   Square 8vo., cloth gilt.   Rs. 4.
"He presents the various sorts and conditions of humanity that comprise
the round of life in Bengal in a series of vivid vignettes. . . . . He
writes with scholarly directness and finish."—*Saturday Review.*

"A pleasant book to read."—*Suffolk Chronicle.*

"The style is pretty pleasant, and the verses run smooth and melodious."—
Indian Mail.

Landholding; and the Relation of Landlord and

Tenant in Various Countries of the World. By C. D. FIELD, M.A., LL.D. 8vo., cloth. Rs. 17-12. (36s.)

"The latter half of this bulky volume is devoted to an exhaustive description and examination of the various systems of Land Tenure that have existed or which now exist in British India. . . . We may take it that as regards Indian laws and customs Mr Field shows himself to be at once an able and skilled authority. In order, however, to render his work more complete, he has compiled, chiefly from Blue-books and similar public sources, a mass of information having reference to the land laws of most European countries, of the United States of America, and our Australasian colonies. . . . The points of comparison between the systems of land tenure existing up till recently in Ireland, and the system of land tenure introduced into India by the English under a mistaken impression as to the relative position of ryots and zemindars, are well brought out by Mr. Field. He indicates clearly the imminence of a Land Question of immense magnitude in India, and indicates pretty plainly his belief that a system of tenancy based on contract is unsuited to the habits of the Indian population, and that it must be abolished in favour of a system the main features of which would be fixty of tenure and judicial rents."—*Field.*

"A work such as this was urgently required at the present junction of discussion upon the landowning question. Mr. Justice Field has treated his subject with judicial impartiality, and his style of writing is powerful and perspicuous."—*Notes and Queries.*

"Mr. Justice Field's new work on 'LANDHOLDING, AND THE RELATION OF LANDLORD AND TENANT IN VARIOUS COUNTRIES,' supplies a want much felt by the leading public men in Bengal. . . . He gives a complete account of the agrarian question in Ireland up to the present day, which is the best thing on the subject we have hitherto seen. Then he has chapters as to the Roman law, the Feudal system, English law, Prussian, French, German, Belgium, Dutch, Danish, Swedish, Swiss, Austrian, Italian, Greek, Spanish, Portuguese, Russian and Turkish land laws, which . . . will enable controversialists to appear omniscient. On the Indian law he tells us all that is known in Bengal or applicable in this province."—*Friend of India and Statesman.*

Banting in India, with some Remarks on Diet

and Things in General. By Surgeon-Major JOSHUA DUKE. Third Edition. Cloth. Re. 1-8.

Queries at a Mess Table. What shall we Eat?

What shall we Drink? By Surgeon-Major JOSHUA DUKE. Fcap. 8vo., cloth, gilt. Rs. 2-4.

Culinary Jottings. A Treatise in Thirty Chapters, on

Reformed Cookery for Anglo-Indian Exiles. Based upon Modern English and Continental principles. With thirty Menus of Little Dinners worked out in detail, and an Essay on our kitchens in India. By "WYVERN." 8vo., cloth. Rs. 5-8.

A Text-Book of Indian Botany, Morphological,
Physiological, and Systematic. Profusely Illustrated. By
W. H. GREGG, B.M.S., Lecturer on Botany at the Hugli
Government College. Cr. 8vo. Rs. 5; interleaved, Rs. 5-8.

Manual of Agriculture for India. By Lt. FREDERICK
POGSON. Illustrated. Crown 8vo., cloth, gilt. Rs. 5
(7s. 6d.)

CONTENTS.—Origin and general character of soils—Ploughing and
Preparing the ground for sowing seed—Manures and Composts—Wheat
cultivation — Barley — Oats — Rye — Rice — Maize — Sugar - producing
Sorghums—Common, or non-sugar-producing Sorghums—Sugar-cane Crops
—Oil-seed Crops—Field Pea, Japan Pea, and Bean Crops—Dall, or Pulse
Crops—Root Crops—Cold Spice Crops—Fodder Plants—Water-nut Crops
—Ground-nut Crops—The Rush-nut, *vel* Chufas—Cotton Crops—Tobacco
Crops—Mensuration—Appendix.

Roxburgh's Flora Indica; or, Description of
Indian Plants. Reprinted literatim from Cary's Edition.
8vo., cloth. Rs. 5 (10s. 6d.)

The Future of the Date Palm in India. (Phœnix
Dactylifera.) By E. BONAVIA, M.D., Brigade-Surgeon,
Indian Medical Department. Crown 8vo., cloth. Rs. 2-8.

Kashgaria (Eastern or Chinese Turkestan),
Historical, Geographical, Military, and Industrial. By Col.
KUROPATKIN, Russian Army. Translated by Major GOWAN,
H.M's. Indian Army. 8vo. Rs. 6-8. (10s. 6d.)

Mandalay to Momien: a Narrative of the Two Expedi-
tions to Western China of 1868 and 1875, under Cols.
E. B. Sladen and H. Browne. Three Maps, numerous
Views and Wood-cuts. By JOHN M. D. ANDERSON.
Thick demy 8vo., cloth. Rs. 5.   [1876.

British Burma and its People: being Sketches of
Native Manners, Customs, and Religion. By Capt.
C. J. F. S. FORBES. 8vo., cloth. Rs. 4-2.   [1878.

Myam-Ma: The Home of the Burman. By
TSAYA (Rev. H. POWELL). Crown 8vo. Rs. 2. [May, 1886.

A Critical Exposition of the Popular "Jihad,"
showing that all the Wars of Mohammad were defensive,
and that Aggressive War or Compulsory Conversion is not
allowed in the Koran, &c. By Moulavi CHERAGH ALI,
Author of "Reforms under Moslem Rule," "Hyderabad
under Sir Salar Jung." 8vo. Rs. 6.

Ancient India as described by Ptolemy: Being a Translation of the Chapters on India and on Central and Eastern Asia in the Treatise on Geography by Klaudios Ptolemaios, the celebrated Astronomer : with Introduction, Commentary, Map of India according to Ptolemy, and a very copious Index. By J. W. McCRINDLE, M.A. 8vo., cloth, lettered. Rs. 4-4.

The Life of H.M. Queen Victoria, Empress of India. By JOHN J. POOL, Editor, "Indian Missionary." With an Original Portrait from a Wax Medallion by Signor C. Moscatti, Assistant Engraver, Her Majesty's Mint, Calcutta. Crown 8vo. Paper, Re. 1. Cloth, Re. 1-4.

From the City of Palaces to Ultima Thule. By H. K. GORDON, with a Map.

Poppied Sleep. By Mrs. H. A. FLETCHER, Author of "Here's Rue for You."

The Bengal Medical Service, April, 1885. Compiled by G. F. A. HARRIS, Surgeon, Bengal Medical Service. Royal 8vo. Rs. 2.
A Gradation List giving Medical and Surgical Degrees and Diplomas, and Universities, Colleges, Hospitals, and War Services, etc., etc.

Ague; or Intermittent Fever. By M. D. O'CONNELL, M.D. 8vo., sewed. Rs. 2.

Book of Indian Eras.—With Tables for calculating Indian Dates. By ALEXANDER CUNNINGHAM, C.S.I., C.I.E., Major-Genl., R.E., Bengal. Royal 8vo., cloth. Rs. 12.

Protestant Missions.—The Fourth Decennial Statistical Tables of Protestant Missions in India, Ceylon, and Burmah. Prepared, on information collected at the close of 1881, by the Rev. J. HECTOR, M.A., Free Church of Scotland ; the Rev. H. P. PARKER, M.A., Church Missionary Society, and the Rev. J. E. PAYNE, London Missionary Society, at the request of the Calcutta Missionary Conference, and with the concurrence of the Madras and Bombay Missionary Conferences. Super-Royal 8vo. Rs. 2-8.

A Map of the Civil Divisions in India, including Governments, Divisions, and Districts, Political Agencies and Native States. Folded. Rs. 1.

The Laws of Wealth. By HORACE BELL. Third Edition. Fcap. 8vo. 8 Ans.

Calcutta to Liverpool by China, Japan, and America, in 1877. By Lieut.-General Sir HENRY NORMAN. Second Edition. Fcap. 8vo., cloth. Rs. 2-8. (3s. 6d.) The only book published on this interesting route between India and England.

Guide to Masuri, Landaur, Dehra Dun, and the Hills North of Dehra; including Routes to the Snows and other places of note; with Chapters on Garhwal (Tehri), Hardwar, Rurki, and Chakrata. By JOHN NORTHAM. Rs. 2-8.

A Handbook for Visitors to Agra and its Neigh-bourhood. By H. G. KEENE, C.S. Fourth Edition. Revised. Maps, Plans, &c. Fcap. 8vo., cloth. Rs. 2-8.

A Handbook for Visitors to Delhi and its Neigh-bourhood. By H. G. KEENE, C.S. Third Edition. Maps. Fcap. 8vo., cloth. Rs. 2-8.

A Handbook for Visitors to Allahabad, Cawn-pore, and Lucknow. By H. G. KEENE, C.S. Second Edition, re-written and enlarged. Fcap. 8vo. Rs. 2-8.

Hills beyond Simla. Three Months' Tour from Simla, through Bussahir, Kunowar, and Spiti, to Lahoul. ("In the Footsteps of the Few.") By Mrs. J. C. MURRAY-AYNSLEY. Crown 8vo, cloth. Rs. 3.

Son Gruel; or, What he met i' the Mofussil (after two Noble Lords). Cantos I and II. Fcap. 8vo. Re. 1 each.

An Historical Account of the Calcutta Collec-torate. From the days of the Zemindars to the present time. By R. C. STERNDALE, author of "Municipal Work in India." 8vo., cloth. Rs. 2.

Departmental Ditties and other Verses. By RUDYARD KEPLING. Second Edition. With additional Verses. Imp. 8vo. Re. 1-8.

Life: An Explanation of it. By W. SEDGWICK, Major, R.E. Crown 8vo., cloth. Rs. 2.

Elementary Statics and Dynamics. By W. N. BOUTFLOWER, B.A., late Scholar of St. John's College, Cambridge, and Professor of Mathematics, Muir Central College, Allahabad. Crown 8vo. Rs. 3-8.

The Landmarks of Snake Poison Literature.
By VINCENT RICHARDS, F.R.C.S. Crown 8vo. Second
Edition. Rs. 2-8.

A Key to the Entrance Course, 1888. Palgrave's
Student's Lyrics and a A Book of Worthies. As Selected
for the Calcutta University Entrance Examination, 1888.
By F. J. ROWE, M.A., Professor, Presidency College.
Fcap. 8vo., 314 pp. Rs. 2-4 ; post free, Rs. 2-6

Indian-English and Indian Character. By ELLIS
UNDERWOOD. Fcap 8vo. Rs. 1.

The Trial of Maharaja Nanda Kumar. A Narrative
of a Judicial Murder. By H. BEVERIDGE, C.S. 8vo.,
cloth. Rs. 10.

Cherry Blossoms : A Volume of Poetry. By GREECE
CH. DUTT. Crown 8vo., 6s.

The Indian Tribute and the Loss by Exchange.
An Essay on the Depreciation of Indian Commodities in
England, &c., and the utter failure of Bimetallism as a
remedy for India's growing burden. By THOMAS INWOOD
POLLARD, author of "Gold and Silver Weighed in the
Balance." Crown 8vo., cloth. Rs. 2-8.

Gold and Silver weighed in the Balance : A
measure of their value ; an essay on wealth and its distribu-
tions during fluctuations in the value of Gold and Silver.
By THOMAS INWOOD POLLARD, author of "The Indian
Tribute, &c." Crown 8vo., cloth. Rs. 2-8.

Seonee : or, Camp Life on the Satpura Range. A Tale of
Indian Adventure. By R. A. STERNDALE, Author of
"Mammalia of India," "Denizens of the Jungles." Illustrated by the Author. With an Appendix containing an
account of the District of Seonee in the Central Provinces
of India. Second and cheaper edition, post 8vo. Rs. 6.
(8s. 6d.)

Soundness and Age of Horses. With one hundred
illustrations. A Complete Guide to all those features
which require attention when purchasing Horses, distin-
guishing mere defects from the symptoms of unsound-
ness, with explicit instructions how to conduct an examination of the various parts. By Capt. M. H. HAYES.
Post 8vo. Rs. 6 (8s. 6d.)

WORKS IN THE PRESS.

On Horse Breaking. By Capt. M. H. HAYES. Numerous
Illustrations by J. H. OSWALD BROWN. Square.

1. Theory of Horse Breaking. 2. Principles of Mounting.
3. Horse Control. 4. Rendering Docile. 5. Giving Good Mouths.
6. Teaching to Jump. 7. Mount for First Time. 8. Breaking for
Ladies' Riding. 9. Breaking to Harness. 10. Faults of Mouth.
11. Nervousness and Impatience. 12. Jibbing. 13. Jumping
Faults. 14. Faults in Harness. 15. Aggressiveness. 16. Riding
and Driving Newly-Broken Horse. 17. Stable Vices.

The Points of the Horse. A Familiar Treatise on Equine
Conformation. By Capt. M. H. HAYES. Illustrated by
J. H. OSWALD BROWN. Describing the points in which the
perfection of each class of horses consists; illustrated by very
numerous reproductions of Photographs of Living Typical
Animals: forming an invaluable guide to owners of horses.

Echoes from Old Calcutta: being chiefly Reminiscences
of the days of Warren Hastings, Francis, and Impey. By
H. E. BUSTEED. Second Edition. Illustrated.

"Dr. Busteed has made an eminently readable, entertaining, and by no
means uninstructive volume; there is not a dull page in the whole book."
—*Saturday Review.*

"The book will be read by all interested in India."—*Army and Navy
Magazine.*

Hand-Book to the Drill in "Extended Order."
Part III. Field Exercise. 1884. With Plates.

The Culture and Manufacture of Indigo, with a
Description of a Planter's Life and Resources. By WALTER
MACLAGAN REID. Crown 8vo. With twenty full-page Illus-
trations.

"It is proposed in the following Sketches of Indigo Life in Tirhoot and
Lower Bengal to give those who have never witnessed the manufacture of
Indigo, or seen an Indigo Factory in this country, an idea of how the finished
marketable article is produced : together with other phases and incidents
of an Indigo Planter's life, such as may be interesting and amusing to
friends at home."—*Introduction.*

Firminger's Manual of Gardening for India.
A New Edition, thoroughly Revised and Re-written. With
many Illustrations. By J. H. JACKSON, Editor, *Indian
Agriculturist.*

Ince's Guide to Kashmir. Revised and Re-written. By
Surgeon-Major JOSHUA DUKE.

Game, Shore, and Water Birds of India. By Col.
A. LE MESSURIER, R.E., with 111 Illustrations. A *vade
mecum* for Sportsmen.

LAW PUBLICATIONS.

Manual of Revenue and Collectorate Law: with Important Rulings and Annotations. By H. A. D. PHILLIPS, Bengal Civil Service. Crown 8vo. cloth. Rs. 10. (21s.)

> CONTENTS:—Alluvion and Diluvion, Certificate, Cesses, Road and Public Works, Collectors, Assistant Collectors, Drainage, Embankments, Evidence, Excise, Lakhiraj Grants and Service Tenures, Land Acquisition, Land Registration, Legal Practitioners, License Tax, Limitation, Mortgages, Opium, Partition, Public Demands Recovery, Putni Sales, Registration, Revenue Sales, Salt, Settlement, Stamps, Survey, and Wards.

The Negotiable Instruments Act, 1881: being an Act to define and amend the Law relating to Promissory Notes, Bills of Exchange and Cheques. Edited by M. D. CHALMERS, M.A., Barrister-at-Law, Author of "A Digest of the Law of Bills of Exchange," &c., and Editor of Wilson's "Judicature Acts." 8vo., cloth. Rs. 7. (10s. 6d.)

A Commentary on Hindu Law of Inheritance, Succession, Partition, Adoption, Marriage, Stridhan, and Testamentary Disposition. By Pundit JOGENDRO NATH BHATTACHARJI SMARTA SIROMANI, M.A., D.L. Demy 8vo. Price Rs. 12, cloth, gilt.

"All the important questions of Hindu Law are discussed in this work in accordance with those rules and principles which are recognized among Hindu jurists as beyond dispute. By going through the work the reader will become familiar with the Hindu lawyers' modes of thought and reasoning, and will be prepared to argue or discuss any point of Hindu Law.

"Babu Bhattacharji is the greatest name in the recent history of the University. He has already made his mark, having written a really original work on Hindu Law, which must assert itself against the crude compilations and false views of European writers."—*Reis and Rayyat,* December 26th, 1885.

The Indian Limitation Act; Act XV. of 1877. (As amended by Act XII. of 1879, and subsequent enactments), with Notes. By H. T. RIVAZ, Barrister-at-Law, Advocate, N.-W.-P., and Punjab. Third Edition. Royal 8vo., cloth. Rs. 10.

A Chaukidari Manual; being Act VI. (B.C.) of 1870, as amended by Acts 1. (B.C.) of 1871 and 1886. With Notes, Rules, Government Orders, and Inspection Notes. By G. TOYNBEE, CS, Magistrate of Hooghly. Crown 8vo. cloth. R. 1.

xxii   *Thacker, Spink & Co., Calcutta.*

Manual of the Revenue Sale Law and Certificate

Procedure of Lower Bengal, being Act XI. of 1859; Act VII. (B.C.) of 1868; and Act VII. (B.C.) of 1880 : The Public Demands Recovery Act, including Selections from the Rules and Circular Orders of the Board of Revenue. With Notes. By W. H. GRIMLEY, B.A., C.S. 8vo. Rs. 5-8 ; interleaved, Rs. 6.

The North-Western Provinces' Rent Act, being

Act XII. of 1881, as amended by Act. XIV. of 1886. With Notes, &c. By H. W. REYNOLDS, C.S. Demy 8vo., cloth. Rs. 7.

The Bengal Tenancy Act. Being Act VIII. of 1885.

With Notes and Annotations, Judicial Rulings, and the Rules framed by the Local Government and the High Court under the Act. For the guidance of Revenue Officers and the Civil Courts. By M. FINUCANE, M.A., C.S., Director of the Agricultural Department, Government of Bengal, and R. F. RAMPINI, M.A., C.S., Barrister-at-Law, District and Session Judge. Second Edition. [In the Press.

The Inland Emigration Act ; with Orders by the

Lieutenant-Governor of Bengal ; Forms by Government of Bengal ; Resolution of the Government of India ; Resolution of the Government of Assam ; Rules made by the Chief Commissioner of Assam, and Orders by the Lieutenant-Governor, N. W. P. Interpaged with blank pages for notes. Crown 8vo. Rs. 2-4.

The Hindu Law of Inheritance, Partition, and

Adoption according to the Smritis. By Dr. JULIUS JOLLY, Tagore Law Lecturer, 1883. Rs. 10.

The Bengal Local Self-Government Act (B.C.

Act III of 1885), and the general Rules framed thereunder. With Critical and Explanatory Notes, Hints regarding Procedure, and reference to the Leading Cases on the Law relating to Local Authorities. To which is added an Appendix containing the principal Acts referred to, &c., &c. ; and a Full Index. By F. R. STANLEY COLLIER, B.C,S., Editor of " The Bengal Municipal Act." Crown 8vo. Rs. 4.

An Income Tax Manual, being Act II. of 1886. The

Rules, Rulings and Precedents, &c,, and Notes. By W. H. GRIMLEY, B.A., C.S., Commissioner of Income Tax, Bengal. Royal 8vo. Rs. 3-8 ; interleaved, Rs. 4.

The Pocket Penal, Criminal Procedure and Police
Codes; also the Whipping Act and the Railway Servants' Act. With General Index. 1 Vol. Rs. 4.

The Pocket Civil Procedure Code, with Court Fee,
Indian Evidence, Specific Relief, Indian Registration, Limitation, and Stamp Acts. With General Index. 1 Vol. Rs. 4.

The Indian Penal Code and other Laws and Acts
of Parliament relating to the Criminal Courts of India. With Notes. By J. O'KINEALY, Judge of the High Court, Calcutta. Third Edition. Royal 8vo. Rs. 12.

Legislative Acts of the Governor General of India in
Council; published annually with Index. Royal 8vo., cloth. 1872, Rs. 10; 1873, 1874, and 1875, Rs. 5 each; 1876, Rs. 6; 1877, Rs. 10; 1878, Rs. 5; 1879, Rs. 5; 1880, Rs. 4; 1881, Rs. 8; 1882, Rs. 15-8; 1883, Rs. 5; 1884, Rs. 5; 1885, Rs. 5.; 1886, Rs. 5s.

Indian Case-Law on Torts.—By R. D. ALEXANDER,
Bengal Civil Service. Crown 8vo., cloth. Rs. 4.

Introduction to the Regulations of the Bengal
Code. By. C. D. FIELD, M.A., LL.D. (specially reprinted for the use of students, etc.). In crown 8vo., cloth. Rs. 3.

The Law of Evidence in British India. By C. D.
FIELD, M.A., LL.D, Judge of the High Court, Calcutta. Fourth Edition Rs. 18.

The Indian Limitation Act XV. of 1887. Edited
with Notes of Cases, by R. D. ALEXANDER, C.S., Judge, Allahabad. Crown 8vo. Rs. 2-4.

The Indian Contract Act No. IX. of 1872. To-
gether with an Introduction and Explanatory Notes, Table of Contents, Appendix, &c. By H. S. CUNNINGHAM, M.A., one of the Judges of H.M.'s High Court of Judicature, Calcutta; and H. H. SHEPHARD, M.A., Barrister-at-Law. Fifth Edition.

The Practice of the Presidency Court of Small
Causes of Calcutta. The Presidency Small Cause Courts Act (XV. of 1882), with Copious Notes; the Code of Civil Procedure, with Notes and References; the Rules of Practice, Institution, and Court Fees; and a complete Index. By R. S. T. MacEWEN, Barrister-at-Law, one of the Judges of the Presidency Court of Small Causes of Calcutta. Thick 8vo. Rs. 11. Cash 10.

The Code of Criminal Procedure. Together with
Rulings, Circular Orders, Notifications, &c., of all the High Courts in
India, and Notifications and Orders of the Government of India and
the Local Governments. Edited, with Copious Notes and full Index,
by W. F. AGNEW, and GILBERT S. HENDERSON, M.A., Barristers-at-
Law. Second Edition. Royal 8vo., cloth, Rs. 18.

The Law of Specific Relief in India; being a Com-
mentary on Act I. of 1877. By CHARLES COLLETT, late of the Madras
Civil Service, of Lincoln's Inn, Barrister-at-Law, and formerly a Judge
of the High Court at Madras. Demy 8vo. Rs. 10. Cash 9. (14s.)

The Law of Intestate and Testamentary Suc-
cession in India; including the Indian Succession Act, &c., with a
Commentary. With Notes and Cross References. By GILBERT S.
HENDERSON, M.A., Barrister-at-Law. Royal 8vo. Rs. 16.

Manual of Indian Criminal Law: being the Penal
Code, Criminal Procedure Code, Evidence, Whipping,
General Clauses, Police, &c., Acts, with Penal Clauses
of Legal Practitioners' Act, Registration, Arms, Stamp,
&c., Acts. Fully Annotated, and containing all applicable
Rulings of all High Courts arranged under the appropriate
Sections up to date. By H. A. D. PHILLIPS. Thick
crown 8vo. New Edition. Rs. 10.

The Stamp Law of British India, as constituted by
the Indian Stamp Act (I. of 1879). Rulings and Circular
Orders of the four High Courts; Notifications; Resolutions;
Rules; and Orders of the Government of India and of the
various Local Governments; together with Schedules of all
the stamp duties chargeable on Instruments in India from
the earliest times. Edited, with Notes and Index, by
WALTER R. DONOGH, M.A., of the Inner Temple, Barrister-
at-Law. Demy 8vo. Rs. 8.

Code of Civil Procedure (Act XIV. of 1882). With
Notes, &c. By J. O'KINEALY, C.S., Judge of the High
Court, Calcutta. Second Edition, Royal 8vo. Rs. 16.

Law of Intestate and Testamentary Succession
in India, including the Indian Succession Act (x. of 1865),
with a Commentary, and the Parsee Succession Act,
Hindu Wills Act, Probate and Administration Act, Dis-
trict Delegates Act, Acts xii. and xiii. of 1855, Regimental
Debts Acts, Acts relating to the Administrator-General
Certificate Act, and Oudh Estates Act, with Notes and
Cross References and a General Index. By GILBERT S.
HENDERSON, M.A., Barrister-at-Law. Rs. 16.

Comparative Criminal Jurisprudence, being a synopsis of the law, procedure, and case law of other countries, arranged as far as possible under the corresponding sections of the Indian Codes. By H. A. D. PHILLIPS.

Vol. I. Crimes and Punishments. Vol. II. Procedure and Police.

This work will include extracts from the Penal and Criminal Procedure Codes of the State of New York, of Louisiana, of France, Belgium, and Germany, the English statute-law and case-law (up to date), as well as the most important decisions of the Courts of various American States, the Supreme Court of the United States, and the Court of Cassation in Paris ; also extracts from the best works on criminal law and jurisprudence.

[In the Press.

The Indian Law Examination Manual. — By FENDALL CURRIE, Esq., of Lincoln's Inn, Barrister-at-Law. Third Edition. Demy 8vo. Rs. 5.

CONTENTS :—Introduction—Hindoo Law — Mahomedan Law—Indian Penal Code—Code of Civil Procedure—Evidence Act—Limitation Act—Succession Act—Contract—Registration Act—Stamp Acts and Court Fees —Mortgage—Code of Criminal Procedure—The Easement Act—The Trust Act—The Transfer of Property Act—The Negotiable Instruments Act.

The Bengal Municipal Manual, containing the Municipal Act (B. C. Act III. of 1884) and other Laws relating to Municipalities in Bengal, with the Rules and Circulars issued by the Local Government, and Notes. Second Edition, Revised and Enlarged. By F. R. STANLEY COLLIER, B.C.S. Crown 8vo., cloth. Rs. 5.

The Law of Mortgage in India, including the Transfer of Property, with Notes of Decided Cases. The Second Edition of the Tagore Law Lectures, 1876. Revised and partly rewritten. By RASHBEHARY GHOSE, M.A., D.L.

[In the Press.

TAGORE LAW LECTURES.

The Hindu Law; being a Treatise on the Law administered exclusively to Hindus by the British Courts in India. (1870 and 1871.) By HERBERT COWELL. Royal 8vo., 2 vols., cloth, each Rs. 8.

History and Constitution of the Courts and Legislative Authorities. (1872.) By HERBERT COWELL. New Edition. (1884). Demy 8vo. Rs. 6.

Mahomedan Law. By SHAMA CHURN SIRCAR.
Digest of Laws according to Sunni Code. Rs. 9. (1873.)
Sunni Code in part and Imamyah Code. Rs. 9. (1874.)

The Law relating to the Land Tenures of Lower Bengal. (1875.) By ARTHUR PHILLIPS. Rs. 10.

The Law relating to Mortgage in India. (1876.) By RASH BEHARI GHOSE.

The Law relating to Minors in Bengal. (1877.) By E. J. TREVELYAN. Royal 8vo., cloth. Rs. 10.

The Hindu Law of Marriage and Stridhana. (1878.) By GOOROO DOSS BANERJEE. Royal 8vo. Rs. 10.

The Law relating to the Hindu Widow. By TRAILOKYANATH MITTRA, M.A., D.L. Rs. 10. (1879.)

The Principles of the Hindu Law of Inheritance. By RAJCOOMAR SARVADHICARI, B.L. Rs. 16. (1880.)

The Law of Trusts in British India. By W. F. AGNEW, Esq. Rs. 12. (1881).

The Law of Limitation and Prescription in British India. By OPENDRA NATH MITTER. (1882.)

The Hindu Law of Inheritance, Partition, and Adoption, according to the Smritis. By Dr. JULIUS JOLLY (1883.) Rs. 10.

The Law relating to Gifts, Trusts, and Testa- mentary Dispositions among the Mahomedans. By SYED AMEER ALI. (1884.) Rs. 12.

THE INDIAN MEDICAL GAZETTE.

A Record of Medicine, Surgery, and Public Health, and of General Medical Intelligence, Indian and European. Edited by K. McLeod, M.D.

Published Monthly. Subscriptions Rs. 18 per Annum, including Postage.

The Indian Medical Gazette has for more than twenty years earned for itself a growing and world-wide reputation by its solid contributions to Tropical Medicine and Surgery. It is the **Sole** representative medium for recording the work and experience of the Medical Profession in India ; and its very numerous **Exchanges** with all the leading Medical Journals in Great Britain and America enable it not only to diffuse this information broadcast throughout the world, but also to cull for its Indian readers, from an unusual variety of sources, all information which has any practical bearing on medical works in India.

The Indian Medical Gazette is indispensable to every Member of the Medical Profession in India who wishes to keep himself abreast of medical progress, for it brings together and fixes the very special knowledge which is only to be obtained by long experience and close observation in India. In this way it constitutes itself a record of permanent value for reference, and a journal which ought to be in the library of every medical man in India or connected with that country.

The *Gazette* covers altogether different ground from *The Lancet* and *British Medical Journal,* and in no way competes with these for general information, although it chronicles the most important items of European Medical Intelligence. The whole aim of the *Gazette* is to make itself of special use and value to Medical Officers in India, and to assist and support them in the performance of their difficult duties.

It is specially devoted to the best interests of **The Medical Services,** and its long-established reputation and authority enable it to command serious attention in the advocacy of any desirable reform or substantial grievance.

The **Contributors** to *The Indian Medical Gazette* comprise the most eminent and representative men in the profession. and the contents form a storehouse of information on tropical diseases which would otherwise be lost to the world.

UNIFORM SERIES OF ILLUSTRATED WORKS.
Square Imperial 16mo.

FINELY PRINTED AND HANDSOMELY BOUND.

Riding : on the Flat and Across Country. A Guide
to Practical Horsemanship. By Capt. M. H. HAYES. 53
Illustrations by Sturgess and others. Second Edition.
10s. 6d. Rs. 7.

Riding for Ladies: with Hints on the Stable. By Mrs.
POWER O'DONOGHUE. With 91 Illustrations by A. Chantrey
Corbould. 12s. 6d. Rs. 10.

A Natural History of the Mammalia of India,
Burmah and Ceylon. By R. A. STERNDALE, F.R.G.S.,
F.Z.S., &c. With 170 Illustrations by the Author and
others. 12s. 6d. Rs. 10.

The Tribes on My Frontier: an Indian Naturalist's
Frontier Policy. By EHA. With 50 Illustrations by
F. C. Macrae. Third Edition. 8s. 6d. Rs. 5-8.
Most graphically and humorously describes the surroundings of a country
bungalow. The twenty chapters embrace a year's experiences, and provide
endless sources of amusement and suggestion.

Lays of Ind. By ALIPH CHEEM. Comic, Satirical, and
Descriptive Poems illustrative of Anglo-Indian Life. With
70 Illustrations. Seventh Edition. 10s. 6d. Rs. 7.

Wilkins.—Hindu Mythology: Vedic and Puranic. By
Rev. W. J. WILKINS, of the London Missionary Society,
Calcutta. Illustrated by very numerous Engravings from
Drawings by Native Artists. Uniform with "Lays of
Ind," "Riding," &c. 10s. 6d. Rs. 7.

Beddome.—A Popular Handbook of Indian Ferns. By
Colonel R. H. BEDDOME, Author of the "Ferns of British
India," "The Ferns of Southern India." 300 Illustrations
by the Author. Uniform with "Lays of Ind," "Hindu
Mythology," "Riding," "Natural History of the Mammalia
of India," &c. Imperial 16mo. 12s. 6d. Rs. 10.

INDEX TO LAW BOOKS.

INDEX TO GENERAL PUBLICATIONS.

Index to General Publications—*continued.*